31-/-

Basic Principles
of the Tracer Method

BASIC PRINCIPLES OF THE
TRACER METHOD

Introduction to Mathematical Tracer Kinetics

C. W. SHEPPARD

Professor of Physiology
University of Tennessee Medical Units
Memphis, Tennessee

Supported by a U.S. Atomic Energy Commission
Research Contract

John Wiley & Sons, Inc., New York · London

Library of Congress Catalog Card Number: 62–8788
Printed in the United States of America

176984

To Paul F. Hahn—*"another pinch."*

Preface

A quarter of a century of research with radioactive isotopic tracers has given ample testimony to the value of this research tool, and indeed to the tracer method in general, whether it involves radioactive isotopes or not. The tool is admittedly a dull one on occasion, but we are definitely committed to it and to the task of its improvement. We should not regard the situation too pessimistically. It is a characteristic of research techniques that they tend to improve with time in a fashion that cannot always be readily anticipated. Meanwhile our libraries contain many physical, chemical, engineering, and biological tracer studies, but relatively few papers concern themselves with basic principles of the method.

It is the purpose of this book to collate and unify the literature on this subject. Perhaps, as a result of this effort, some existing barriers between fields will be overcome with improved communication between workers. We may also hope that a broader appreciation of the nature of the tracer method and a clearer understanding of its present scope and limitations will be created. In the past, workers who use tracers have generally considered the principles of the method more or less intuitively. The need to crystallize these intuitive concepts is increasingly obvious. Since basic physical principles are being investigated, the familiar mathematical tools of physics must be used, but, because of the special needs of the biological worker, the mathematics will not be rigorous. I beg forgiveness on the one side for too much mathematics and on the other for too little.

In the mathematical field of digital computation methods, we must continue to be prepared for rapid obsolescence. As one example, during the preparation of Chapter 7, a revised FOR TRANSIT system appeared, and the programming instructions had to be altered accordingly. The IBM 650 may soon cease to be the machine of choice, but it would seem that, for more modern IBM machines at least, some of the basic principles will carry over into the new systems. There is also rapid growth in the field of membrane kinetics. Because of the evolution in current ideas our discussion must be supplemented by further literature research.

At an early stage in the writing of this book, I decided to focus attention on central unification rather than peripheral detail. Therefore this book will not, in several specialized instances, provide a substitute for thorough literature reading. Readers will be aided by the bibliography which appears at the end of the book. Considerable effort has been put forth in searching the literature. Nevertheless, the fact that it is practically impossible to screen the enormous number of current American and foreign scientific publications completely means that a few omissions will certainly occur. For this, I offer deepest apologies.

Thanks are due to Corinne Ridolphi for checking the manuscript and suggesting improvements in the presentation. Constructive criticism was provided by many colleagues, in particular, Drs. J. S. Robertson, Gerald A. Wrenshall, and Mones Berman. The illustrations are the work of Dr. James M. Austin, and the mechanics of manuscript preparation were accomplished through the conscientious work of Mrs. Donna K. Simmons. Michael B. Uffer and Don P. Engelberg provided several of the demonstration simulations with the analog computer. Marshall P. Jones prepared the table of the random walk function. Thanks are particularly due to Mr. Owen Bruce and Mr. E. O. Godbold of the Service Bureau Corporation for instruction in the finer points of the operation of the IBM 602A and 650 computing machines respectively. The ideas in this book have benefited by discussions with many colleagues. In particular Dr. L. J. Savage, Dr. A. S. Householder and Dr. John L. Stephenson have constructively influenced the author's thinking. If Chapter 6 possesses any clarity it is because of the stimulus provided from discussions with Dr. Lester Van Middlesworth and Dr. David L. Yudilevich. I assume sole responsibility for any errors which may later be found in these pages.

The manuscript of this book was prepared during the tenure of a Senior Fellowship Grant (SF-152) from the United States Public Health Service.

C. W. SHEPPARD

Memphis, Tennessee
October, 1961

Contents

ix

Selected Symbols
and Abbreviations

Symbol	Page first used	Meaning
S	1	total amount of substance to be traced
ρ	10	rate of exchange of S between 2 compartments
t	10	time
Δ_{12}	12	$a_1 - a_2$
$V_{1,2}$	14	voltage of capacitor 1, 2, used in simulation
$C_{1,2}$	14	capacitance of capacitor 1, 2, used in simulation
$1/R$	14	conductance of resistor used in simulation
S_i	18	amount of S in compartment i
r	18	superscript r denoting one of a number of species of S used as tracers
n	18	maximum value of i, or index of last compartment in a series
^rA_i	19	fractional amount (abundance ratio) of the rth species of S in compartment i
R_i	19	amount of tracer (radioactivity, perhaps) in compartment i
a_i	19	specific activity R_i/S_i in compartment i
$a_i(0)$	19	specific activity in compartment i at zero time
ρ_{ij}	20	rate of transport of S to compartment i from j
$\overset{\circ}{S}_i$	20	abbreviated notation for dS_i/dt
$\rho', \rho'', a',$ etc.	27	primes used to indicate subdivision of a compartment into subsidiary compartments
Δ	34	determinant usually formed from the coefficients of a set of n linear algebraic equations
Δ_x, etc.	34	determinant obtained by replacing the x column in Δ by the column of constants on the right of the linear equations

Symbol	Page first used	Meaning
^rC	39	detector efficiency factor defined as the ratio $^rA_j/^ra_j$
$F(s) = \mathscr{L}[f(t)]$	50	Laplace transform of $f(t)$, i.e., $\displaystyle\int_0^\infty e^{-st} f(t)\, dt$
s	50	Laplace variable equivalent in transform space to t in ordinary space
a_0, a_1, δ, γ	52	general constants normally used in tables of Laplace transforms to indicate relations between functions and their transforms (see Table 1). a_0, etc., are *not* specific activities in this representation
α_i	54	Laplace transform of a_i. In general throughout we will tend to represent transforms by the appropriate greek letter
ρ_i	55	exchange rate in a mammillary system between central compartment and ith peripheral one
β_i	58	exchange rate of the ith peripheral compartment as a fraction of the compartment contents, i.e., ρ_i/S_i
B_i	58	ith peripheral exchange rate as a fraction of central compartmental contents, i.e., ρ_i/S_0
D	58	discriminant of quadratic equation for roots in two compartment systems, i.e., $(\lambda_1 - \lambda_2)^2$
λ_i	60	exponential constant in ith term of a series of exponential functions
$A(s), B(s)$	60	numerator and denominator polynomials of expression to be expanded in partial fractions
X_i	61	coefficient of ith term in a series of exponential functions
w, e, S_c, S_n	66	system parameters used in analysis of the washout and exchange of isotope in Fig. 15
ϵ	72	constant rate of "wash through" of S in a catenary or a cyclic system
ρ_k	72	exchange rate in a catenary system between compartments k and $k + 1$
f_j	72	$\rho_j + \epsilon$
d_j	72	$sS_j + (\rho_{j-1} + \rho_j + \epsilon)$
Δ_j	73	determinant of rank j containing first j rows and columns of Δ
T	85	dummy time-like variable used in convolution integrals
$F_\infty(t)$	86	specific activity of material from infinite compartment
$\Phi_\infty(s)$	86	Laplace transform of $F_\infty(t)$
$[C_i]$	90	concentration of dye in compartment i of a hydrodynamic analog system
$\{V_i\}$	90	volume of compartment i of a hydrodynamic analog system

Symbol	Page first used	Meaning
r_{ij}	91	rate of water movement from compartment j to compartment i in a hydrodynamic analog system
Q_i	95	charge on ith condenser of an electrical analog circuit
R_i	95	resistance of the ith branch of an electrical analog circuit
I_i	95	current in the ith branch
C_i	95	capacitance of condenser i
V_i	95	voltage of condenser i
C_i	101	in Chapter 6, C_i refers to "coefficient unit i"
d**A**	153	vector representing differential element of area (see Fig. 47)
J*	153	vector representing flux of label (see Fig. 47)
$\mathscr{I}$	155	interfusion constant
i, j, k	155	unit vectors parallel to x, y, and z axes, respectively
$\partial/\partial x$, etc.	155	partial derivative with respect to x, etc.
∇	155	gradient operator, e.g., $\partial/\partial x + \partial/\partial y + \partial/\partial z$, etc.
T	157	temperature in thermal analog systems
$[c]$	157	concentration of label at coordinate point x, y, and z
$K(x, t)$	160	general expression for the function in an integral equation known as the "kernel"
$M(t)$	161	amount of a metabolite remaining in a system at time t
$R(t)$	161	rate of renewal of metabolite in a system
$F(t)$	161	fraction of an initial parcel of S introduced at $t = 0$ and remaining until time t
$\bar{t}$	163	arithmetic mean of t taken over the function $F(t)$
s_k	163	net production of the kth species (source strength)
$[c_k]$	163	concentration of the kth species
$J_k, J_k{}^*$	164	components of J and J* parallel to the x axis and perpendicular to a membrane
r	167	radius of a laminar sleeve of flowing liquid
$v(r)$	167	velocity of laminar sleeve of radius r
R, L, V	167	radius length and total volume of cylindrical tube
Q	167	volumetric flow rate through a system
τ	167	fraction of system volume displaced
I	167	amount of label in initial dye bolus
$C(t)$	167	concentration expressed as a function of time
$F(\tau)\,d\tau$	167	fraction of label appearing at the outflow of a system between τ and $\tau + d\tau$
$\delta(t - a)$	172	delta function ("unit spike") at a
σ	189	standard deviation of the normal distribution

Symbol	Page first used	Meaning
κ	196	randomizing constant of a labyrinth
m_i	203	ith moment of a probability density function about the origin
$T_{1,\,2}$, etc.	205	transit time for label through the first, second, etc., branch of a labyrinth
T	206	transit time through any differential branch of a distributed system
f_1, f_2	206	distribution of transit times through the first and second labyrinth of a series system
φ	206	Laplace transforms of $f_1(t), f_2(t)$
$\gamma(s)$	208	Laplace transform of $C(t)$
ρ_m	219	mean fractional exchange rate taken along a tube of flow
ρ, ρ'	219	exchange rates as fractions of internal and external S (throughout Chapter 10, this will be a fractional rate)
p, q	221	natural coordinates, $p = \rho t/v$, $q = \rho'(t - x/v)$
$\Phi(\rho_m)$	225	distribution function of mean exchange rates through a labyrinth
$\psi(t)$	226	equivalent "non-diffusible" concentration relation for diffusible tracer
$G(t)\,e^{\alpha T/\bar{T}}$	226	distribution of transit times for "equivalent non-diffusible" tracer
$\bar{\rho}, \bar{T}$	226	grand arithmetic means of exchange rates and transit times for the entire circulation
$\Lambda_i(t)$	228	localization function of label in the ith branch of a partitioned system
M_i	229	ith moment of $\Phi(\rho_m)$ taken about the mean
$\|r\|, \|a\|$	236	matrix of transport rates and specific activities in multicompartment systems
$\|\mu\|$	238	matrix relating $\|a\|$ and $d\|a\|/dt$ in multicompartment kinetics
$\|X_i\|$	239	matrix of coefficients in matrix formulation of steady-state multicompartment systems
$\|X\|$	241	matrix of fitted coefficients of exponential terms in multicompartment experiments
x_i, y_i	243	x and y coordinates of experimental data to be fitted by least squares
$\hat{y}$	243	fitted value of y
A_n, λ_n	245	parameters to be estimated in least-squares curve fitting
$[n_{ij}]$	247	Gaussian bracket notation to represent sums of cross products in statistical curve fitting

1

Elementary Principles of the Tracer Method

Introduction

The tracer method is a technique for observing a population of specific things such as molecules, living creatures, or other entities by a process of *labeling*.* Observations of the labeled and non-labeled elements as they mingle with one another yield information about the population as a whole. A representative set of elements, the *labeled species*, must be specifically marked in some way. The label thus used may be a radioactive atom, a dye molecule, or any other practical means of identification. The tracer method is often employed in kinetic studies in which, from the movement of the labeled species, the behavior of the entire system is inferred. Most typically the molecules of some *traced substance*, *S*, are labeled by incorporation of isotopically identifiable tracer atoms into their structure. Frequently, rates of movement between two or more *compartments* in a system are estimated from determinations of the compartmental contents as functions of time. Of particular importance is the situation in which a system in a *steady state* is being studied. Here concentrations, or compartmental amounts, may remain constant, although more or less rapid internal *exchange* of material may occur. We need some means of tracing this exchange internally, since we cannot determine it by gross external observations of compartmental contents alone. In some cases, particularly in the field of biology, one or more compartmental systems may be proposed as *theoretical models* for the interpretation of results. Whether compartmental or not, models may represent *closed systems*, which are isolated from their surroundings, as distinct from *open*

---

* In Ch. 1, certain terms having special connotations will be italicized when first mentioned.

I

systems. Although the full appreciation of the basic principles of the method involves considerations in complex systems, most of the concepts in later chapters will be extensions of some rather simple systems which will be discussed in this chapter.

In the familiar use of isotopic tracers, the method does not involve the labeling of one or a few atoms and the tracing of them through a physical, chemical, or biological system. Such individual observations are beyond the capabilities of any telescope or microscope which exists at the present time. As soon as we introduce more than one labeled atom, we must either label each one differently or else at once be resigned to the observation of mean statistical behavior of a population of atoms with a common label. Even in the case of so-called *carrier-free* isotopes, the numbers of tracer atoms can be large enough to eliminate statistical fluctuations. Thus the problem becomes one of diffusion in the more general sense of the term. We are thus subjected to certain special limitations in the knowledge to be obtained from tracer experiments.

Perfect versus imperfect tracers. The total scope of tracer experiments ranges from the most crude to the most refined. What is required depends to a great extent on the problem at hand. In a hydraulic system, how good a tracer for water is dissolved fluorescein? How adequately does a molecule of plasma protein, labeled with the dye T-1824, trace plasma proteins in the body? How well does elaidic acid serve as a chemical tracer for its cis isomer oleic? Are atoms of radioactive iron in a reactor-exposed piston ring adequate to trace iron worn from the ring in service? In no case is there a tracer which can be considered as 100-per-cent-perfect, since the only atoms which will exactly imitate the behavior of the population being traced are those that are identical with the population and thus incapable of separate identification. The fluorescein molecules will not trace water during evaporation. Biological enzyme systems will interact rapidly with one stereoisomer, and slowly or not at all with another. All of these experiments, however, may be idealized in terms of a conceptual model experiment in which labeled and non-labeled species are assumed to behave identically.

At the molecular level, constituent atoms selected as tracers or non-tracers may seem to be nearly the same when their nuclear masses differ by at most a few mass units; nevertheless, some differences can exist (1) and may prove embarrassing at times. The problems are usually most significant when multistep kinetical processes are involved with small molecules in which the small isotopic mass differences are most manifest. The effects are also enhanced for the lighter atoms such as deuterium where increase of one mass unit may involve doubling the mass. In the present treatment we will consider only the ideal model of experiments with *"perfect"* tracers. As the tracer art progresses, some workers have begun to introduce corrections into their data for *isotope effects*.

Tracer statics versus kinetics. The isotopic tracer method is occasionally employed as a substitute for analytical determination of amounts of chemical substances. We may consider as an example a hypothetical experiment in the field of tropical medicine. A comparison of the trypanocidal effect of trivalent versus pentavalent arsenic in the central nervous system might require a knowledge of spinal-fluid levels following administration of a test dose. In terms of a tracer experiment, we might imagine the special preparation of a tracer amount of Tryparsamide® containing As^{76}. A detector of As^{76} might be used to determine the activity of spinal fluid samples. From the activity of a sample and the activity per milligram of the original material, the amount in the sample can be estimated. It is a fact, however, that analytical chemical methods for arsenic are quite sensitive. Furthermore only small corrections for sample blanks are to be expected since normal tissue levels of arsenic are very low. Here then, as a competing method to chemical analysis, the tracer method might generally be of academic interest only.

Radioactive tracers do occasionally provide useful substitutes for chemical analyses. Nevertheless such procedures are not making use of the full possibilities of the tracer method, a method which can, at times, yield information otherwise very difficult to obtain. Suppose, for example, that we wish to determine the total number of fish in a lake. We may capture a small number of the fish and attach metal tags to them. After releasing the labeled specimens and allowing them to mix with the unlabeled members, we assay the fraction of labeled fish in a test sample and are able to estimate the size of the total population which is otherwise inaccessible to observation. When the principle is used—for example, with K^{42} to estimate the total body potassium in man from measurements of total exchangeable K (2)—it is usually called the method of *isotope dilution*. Such experiments can be considered as examples of a branch of isotope technique which we might term "tracer statics." The method, when thus employed, permits the estimation of an amount of material or a number of objects when only a portion of the system may be sampled.

The principal difficulty is insuring uniformity of mixing. How long must we wait for the completion of the process? This question brings us from statics into kinetics. In some instances, kinetic studies are necessary merely to determine mixing times in isotope dilution measurements. At other times, however, kinetic studies may furnish information concerning processes of mixing, *transport rates*, *exchange rates*, and similar matters, which may be of considerable scientific interest. In fact, rates of self-diffusion in physical systems can reliably be estimated only by kinetic studies using perfect tracers.

The proof of the rapidity with which biochemically important molecules in biological steady-state systems exchange is of outstanding interest. The early success of Schoenheimer in this field remains classic, but his demonstrations were scarcely more than qualitative in nature. Quantitation requires the mathematical analysis of linear systems, the use of the equations of diffusion theory, the investigation of the mathematics of probability, and at times a rather sophisticated type of reasoning.

It is unfortunate that the quantitative approach to tracer kinetics has at times been misused and erroneous precision attributed to its results. One of the difficulties, particularly among biological workers, has been the lack of experience with the mathematical methods. A partial solution for this difficulty may be achieved by electronic simulation and analog computation. This will be emphasized in Chapter 6.

Tracers and information.   The physical statistics literature has been concerned with the characterization of a system of molecules or atoms in terms of its content of entropy or of information (3). For illustrative purposes let us consider a human erythrocyte equilibrating in plasma whose potassium contains radioactive K^{42}. At the beginning of the experiment, all of the potassium in the cell is non-radioactive (excluding the natural K^{40}). Thus, the informational content of the system may be qualitatively expressed as the proposition, "none of the K^{42} is in the cell." We can quantitate the informational content of the cell in a way analogous to the measurement of the informational content of a computer register containing binary digits. Initially the non-radioactive condition of the ions is analogous to the condition that follows the resetting of the computer digits to zero. The red blood cell may contain approximately $3\frac{1}{2}$ billion K ions which initially are non-radioactive. Thus at the beginning of the experiment, the cell contains $3\frac{1}{2}$ billion "bits" of information. However, just as the contents of the computer register may deteriorate because of "noise," so may a continuing exchange of potassium between the cell and its surroundings cause information to be lost. This loss is not only in the cell but also in the surrounding suspension medium. At the same time the entropy of the system increases until a state is reached where the specific activity inside and outside the cell are indistinguishable. The loss of information originally placed in the system by labeling is now complete. Since this information may at times have been obtained only at great expense, it is of interest to consider the inefficiency of our present methods for extracting it.

There are, of course, better and poorer methods for preserving information in such situations. If nothing is done to retain the information in a beaker of radioactive KCl, then it will be lost to us if our only method is to study the system with a detector of radioactivity. In this case, then, a few days of decay will suffice to effect the essentially complete degradation of the system. However, if methods were available to assay the beaker for the Ca^{42} which might remain after the radioactive decay, much of this loss could be avoided.

The Use Of Isotopes as Tracers

Isotope mixing, diffusion, and interfusion. Whether isotopes or other labeling agents are employed, the tracer method can be used to study passive diffusion kinetics in many types of systems which are in thermodynamic equilibrium. The isotopic label is a practical advantage in one of the more general situations, such as the study of systems which are neither in thermodynamic equilibrium nor in a steady state. Equations which may be used to describe tracer kinetics must apply to these more general systems. If we can imagine all processes in which organized systems proceed to a greater state of chaos, we can then apply the term "diffusion" in a generalized sense. However, this kind of diffusion represents more than a simple averaging of random walks of molecules as they mix, for example, with a solvent, with molecules of a foreign gas, or with empty space.

For an illustration, we may again return to the erythrocyte, where the process of selective accumulation of potassium has been of interest. Under natural conditions there appears to exist in the cell a set of transport processes maintaining a large potassium concentration gradient between the cell water and the plasma. A steady state exists in which the system remains out of thermodynamic equilibrium. Whatever the processes might be by which K is accumulated in the cell, they act on K^{42} ions as well as on the other K isotopes. The net result is that, starting from an intracellular concentration of zero, K^{42} is accumulated in the normal cell approximately twentyfold in concentration over that of the plasma. This process is in a direction contrary to diffusion as we know it, and yet we may still learn something of individual mean inward and outward transport rates from analysis of rates of K^{42} and total K movement. Since "diffusion" is a somewhat misleading term and suggests passive mixing of solute with solvent, there may be some advantage to the term "interfusion" here (4). This would be the process of mixing a labeled with an non-labeled substance, rather than that of mixing a solute with a solvent. No labeling method other than the isotopic one has been found as yet for studying this type of kinetic process.

Tracers and the exponential function. As an illustration of a typical tracer experiment the study of body water in man can be considered (5). This system provides a good illustration of the use of isotopic labels in particular. The experimental subject ingests, or receives by vein, a minute quantity of "heavy" water containing either deuterium or the radioactive isotope tritium in place of ordinary hydrogen. Usually, the deuterium or tritium oxide will be diluted with a certain amount of "carrier" water. After this heavy water has been given time to mix internally in the body, samples of body fluids are obtained, and the amount of deuterium oxide per gram of water is determined and compared with the amount per gram in the original ingested sample. For example, if half a liter of water was originally ingested, and if, from the concentration in the final sample, a dilution of 1/120 can be computed, then the total amount of water into which the tracer was diluted (including the tracer itself) is evidently $0.5 \times 120 = 60$ liters. The result yields an estimate of the total amount of body water.

After a few hours all readily obtained samples of body fluid contain the same concentration of labeled water within normal experimental limits. At the same time the equilibrated concentration does not change very rapidly; therefore the concept of a true volume of dilution into which labeled water mixes uniformly does not seem difficult to accept. However, there is a turnover of body water which, in most cases, proceeds rather steadily so that ingestion and excretion of water nearly balance. Ignoring small fluctuations, we observe that the system remains not in equilibrium with its environment but in a "steady state," with non-labeled water sweeping through it and washing labeled water out. After 8 to 11 days, an average of approximately half the labeled water has been replaced.

Because of the progressive change in level of isotope in the body, more can be learned about body water than a simple determination of its total amount. If 7 per cent of the activity of the body fluids has disappeared in a day (Fig. 1), then this is the percentage rate at which water is being lost and replaced by non-labeled water. The situation is analogous to that of negative compound interest where interest is compounded continually (6). The level of activity follows the exponential law of decline in which the rate of loss is always proportional to the amount present (Fig. 1A). The ratio of these quantities may be termed the *fractional rate of loss*, and, when multiplied by 100, this ratio becomes the *percentage rate of loss*. Whatever fractional loss occurs over a given period of time, the same fractional decline will occur over any later equal period. Plotting, on

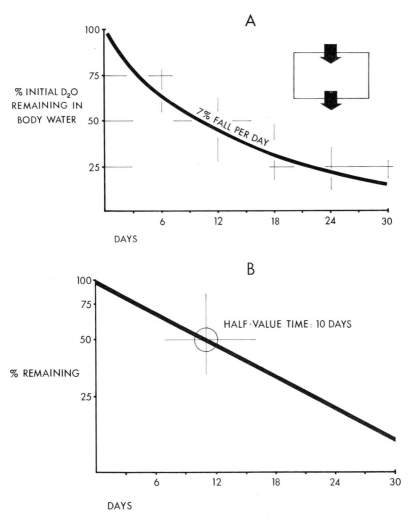

Fig. 1. A–A graph of the decline of the percentage of the initial D_2O remaining in the body water as a function of time in days; B–the plot of the graph illustrated in A when semi-log coordinates are used. The "half-value time" of 10 days is the time required to reduce the percentage of initial label present in the body water from 100% to 50%.

semi-log paper, the level of activity as a function of time will yield a straight line (Fig. 1B). This has been verified for water in man. Since both the label and, thus, the labeled water decline by half in 8 to 11 days, it is said that a "biological half-life" of 8 to 11 days exists for body water.

These experiments can thus be interpreted in terms of the idealized model of washout from a simple one-compartment system. Although water seems to behave this way, the majority of body constituents do not. Often systems of several compartments must be invoked to fit experimental data. It may even be shown, with sufficiently precise methods, that ingested tritium may, in small amounts, exchange with hydrogen in organic combination. If this situation occurs, then the organic tritium is said to be in a separate compartment or pool from water. In general, the model with which results of a tracer experiment are to be compared will depend upon the precision of the method and the desired precision of the conclusions to be drawn. The same may also be said for the validity of biological half-life.

Closed versus Open Systems

Two-compartment systems; initial rates. The most familiar systems encountered in the early literature are those containing two compartments between which material is considered to move. Many

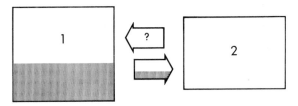

Fig. 2. A two-compartment system in which compartment 1 initially contains the total amount of tracer. The amount of label present is designated by the shaded area. The arrow from left to right may be determined from the percentage of label in compartment 1 and from the rate of disappearance of label from the compartment. Since no label is being carried in the reverse direction the arrow from right to left is not determined by measurements of amount of tracer alone. It may be obtained, however, from the rate of dilution of compartment 1.

analyses of the equations expected in the case of pure exchange can be found (7, 8). Systems are closed if no material can enter or leave and open if communication with the external environment is permitted. Usually, all of the tracer is initially assumed to be in one compartment (Fig. 2). Here the tracer method becomes particularly simple when observations are made early in time. Labeled material is then moving one way and unlabeled material is moving the other way.

Since a general two-compartment system may not be assumed to be in a steady state, two opposing rates must be determined. In Fig. 2 these rates are the opposing arrows. The shading in compartment 1 and in the one arrow represents the fraction of total material which is labeled. If this initial fraction or percentage is known, then from the rate of movement of tracer alone, (that is, its rate of loss from the compartment), the total rate from 1 to 2 is established both for tracer and for traced material. The opposing rate can be obtained by labeling compartment 2 alone in a second identical experiment. If it is necessary to obtain both opposing rates in a single experiment, it is sufficient to determine as functions of time the total amount of labeled and non-labeled material in compartment 1. This determination will yield not only the rate at which material is leaving, but also the overall rate of increase or decrease in the total amount of material in the compartment from which the opposing rate may be obtained by difference. There is still a third way in which the opposing rate could be determined: compartment 2 could be labeled with a second, separately distinguishable label, if one can be used, and the progress of both labels followed in one experiment. One should realize, however, that the absence of label in compartment 2 also serves as a means for tracing its contents. Thus it is that the upper arrow in Fig. 2 may be determined from the initial rate of dilution of the label in compartment 1.

When the system is closed and in a steady state the two rates become a single rate of exchange. In interpreting the results, whether the steady state is known from chemical data or whether it is inherent in the proposed model of the observed system, we are actually providing more information than is determined from a single tracer rate. Of course in a steady-state system, if there is a possibility of more than two compartments, analyses in compartment 1 will yield only the overall outgoing rate which must equal the total incoming rate. If it can be verified that the total content of tracer in both compartments remains constant, it is then clear that the system is closed.

There is an essential difference between the interpretation of initial observations expressed in absolute amounts of a tracer substance, and interpretation of those expressed in fractional amounts. If nothing is entering a compartment, any outgo will produce a decline in the absolute amount of tracer but no change in the fractional or percentage amount, since labeled and non-labeled material are lost together. Changes in fractional amounts represent dilution of tracer by the inflow of non-labeled material. Rates of fractional change are measures only of rates of inflow; therefore, if these rates reflect outflow, it is because outflow and inflow are equal.

It is usually convenient to formulate tracer kinetics in terms of fractional rates or fractional amounts, and, indeed, the absolute amount of a tracer would generally have little experimental significance since it may be arbitrarily selected by the investigator. Fractional amounts of tracer may be expressed directly in terms of *abundance ratios*. In a sample of deuterated water containing 1 per cent of D_2O, this would numerically be the abundance ratio of deuterium. Usually, however, it is not necessary to measure this ratio, but, rather, something proportional to it. When radioactive tracers are used, this ratio is called the *specific activity*; that is, the ratio of the measured tracer radioactivity of a sample, as registered on a radioactivity counting device, to the chemical amount of total traced material in the sample. Thus, for example, a sample of phospholipid which registers 1000 counts per minute and contains 0.2 mg of total phosphorus would have a specific activity of 5000 counts per minute per milligram. The specific activity is not employed when a dye is used as a tracer, but a quantity analogous to it would be the amount in standard arbitrary units of dye per milligram of dyed material. This might be milligrams of dye, or it might be arbitrary units of optical density.

When tracer kinetic equations are formulated in terms of abundance ratios, there is obviously no basis for choosing one isotope over another as the labeling agent. When the specific activity is used, the radioactive isotope is singled out for special attention. Nevertheless it is possible to formulate the principles of tracer kinetics in terms of the idea of "non-labeled" material as the tracer. In the process of tritium labeling in the body water system, for example, it is as meaningful to consider the rate at which non-radioactive water dilutes the radioactive as to consider the rate at which radioactive water leaves the system. It is also possible in principle to use radioactive water to displace water which is initially non-radioactive. Any of these experiments would yield the same result.

Kinetic equations for two compartments. In an exchanging system, soon after label appears in compartment 2, it will begin to return into compartment 1. For pure exchange, formulating this situation mathematically for radioactive tracers, we adopt the following definitions and notation:

S = total amount of traced substance in the system measured in moles, grams, etc.

S_1, S_2 = amounts of S in compartments 1 and 2 respectively.

ρ = rate of exchange between compartments, measured in millimoles per second, grams per hour, etc.

R_1, R_2 = absolute amount of tracer in compartments 1 and 2 measured in counts per minute, microcuries, etc.

a_1, a_2 = specific activities in compartments 1 and 2, i.e., amount of tracer per unit amount of S, measured in counts per minute per gram, microcuries per mole, etc.

t = time in seconds, days, etc.

$a(0)$ = specific activity at zero time in compartment 1.

Figure 3 shows the way in which the changes in the compartments can be expressed. Since the amount of tracer equals the amount per gram of S times the amount of S, we have

$$R_1 = a_1 S_1 \qquad R_2 = a_2 S_2$$

Since S_1 and S_2 are constant, the rates of change of R_1 and R_2 are

$$\frac{dR_1}{dt} = \frac{S_1 \, da_1}{dt} \qquad \frac{dR_2}{dt} = \frac{S_2 \, da_2}{dt}$$

These rates in each compartment are given by the inward rates of transport of R minus the outward rates. The amount of R leaving per

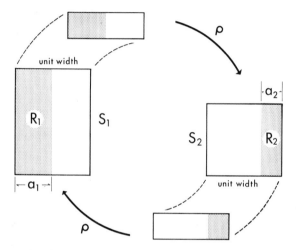

Fig. 3. Schema showing changes in compartments. Compartmental amounts are represented by rectangles of lengths S_1 and S_2, respectively and unit width. If a_1 and a_2, the specific activities, are marked off on the unit sides of the respective compartments, then the amounts of tracer in the two compartments are designated by the shaded areas R_1 and R_2. The smaller rectangles denote the exchange of labeled material between compartments. ρ is the rate of exchange.

second is equal to the amount (say, millimoles) transported per second times the radioactivity per unit amount (say, microcuries per millimole). Similar relations hold for the inward rates. Thus

$$\frac{dR_1}{dt} = \frac{S_1 \, da_1}{dt} = \rho(a_2 - a_1)$$

$$\frac{dR_2}{dt} = \frac{S_2 \, da_2}{dt} = \rho(a_1 - a_2) \qquad (1)$$

We may note that in no case does the volume of a compartment or a concentration occur in these expressions since the quantities are actually irrelevant.

The equations can be simplified if we substitute $\Delta_{12} = a_1 - a_2$ into the system and solve for Δ_{12}, the difference in the specific activities. Thus

$$\frac{da_1}{dt} - \frac{da_2}{dt} = \frac{d\Delta_{12}}{dt} = -\rho\left(\frac{1}{S_1} + \frac{1}{S_2}\right)\Delta_{12}$$

This equation expresses the fact that the rate of change of Δ_{12} is proportional to Δ_{12}. Therefore, the solution yields an exponential relation. Since a_1 ultimately approaches a_2, the terminal value of Δ_{12} approaches zero. If all of the activity is initially in compartment 1, whose specific activity is $a(0)$, then $\Delta_{12}(0) = a(0)$. Thus the solution is

$$\Delta_{12} = (a_1 - a_2) = a(0)e^{-\rho(1/S_1 + 1/S_2)t} \tag{2}$$

Since the total activity in the system is constant,

$$S_1 a_1 + S_2 a_2 = S_1 a(0)$$

Solving this equation simultaneously with eq. 2 for a_1 and a_2 we obtain

$$a_1 = \frac{a(0)}{S}(S_1 + S_2 e^{-\rho St/S_1 S_2})$$

$$a_2 = \frac{a(0)S_1}{S}(1 - e^{-\rho St/S_1 S_2}) \tag{3}$$

Exponential specific activity relations yielding linear semi-log plots have been reported for a number of two-compartment systems (9).

Equations 3 will not yield linear semi-log plots directly, but if the difference between the specific activity and the equilibrium value be plotted, linearization is achieved. A difficulty with this approach is the likelihood that a precise equilibrium value may not be known. Errors in this figure can produce curvature in a linear plot or even tend to linearize a plot which actually shows curvature. The more reliable policy is to plot the values of Δ_{12}.

One-compartment open systems.   The equations for a closed two-compartment system can be used to express the relation in an open one-compartment system. If, for example, S_2 is allowed to approach infinity, this compartment will simulate an infinite reservoir, and the washout process in S_1 will simulate exchange with this reservoir. Thus, if we substitute $a_2 \to 0$ and $1/S_2 \to 0$ in eq. 2, the expression for a_1 becomes

$$a_1 = a(0)e^{-\rho t/S_1} \tag{4}$$

where ρ is the rate with which material is transported through S_1. It may be noted that the kinetic behavior depends only on the ratio of ρ to S_1, not on their individual values.

Solving for rates.   The principal aim of kinetic tracer experiments in compartmental systems is to obtain transport rates. The procedure for the closed two-compartment steady-state system will illustrate general principles in more complex cases. The most straightforward procedure would be to solve directly from the measured values of the S's and a's in eqs. 1. A plot of $a_1(t)$ could be made, for example, and its slope determined. Dividing by the measured value at that instant of $a_2 - a_1$ yields for the fractional exchange rate

$$\frac{\rho}{S_1} = \frac{da_1/dt}{a_2 - a_1} \tag{5}$$

If S_1 is known, then multiplying both sides by S_1 yields the absolute rate ρ directly. It is clear that this computation requires a knowledge of both a's at the time t, and a knowledge of S_1. It may be noted that ρ is an actual exchange rate and not simply a non-specific first-order rate constant. It is *not* a diffusion constant although it may relate to one in special systems where only diffusion is being studied.

When ρ varies with time, eq. 5 still applies, and measurements at different times will yield a ρ whose time variations may then be expressed. If the system is non-steady, a generalization of the method for two rates may be readily constructed, but two equations must be solved for the two unknowns.

If ρ is known to be constant, and if, in the presence of experimental error, maximum precision is desired, it becomes advisable to fit curves to $a_1(t)$ and $a_2(t)$. Here it is convenient to use the integrated relation 2. Taking logs of both sides,

$$\log (a_1 - a_2) = \log a(0) - \rho\left(\frac{1}{S_1} + \frac{1}{S_2}\right)t \tag{6}$$

we obtain an expression which is linear in t. From this, if the assumptions are correct, a semi-log plot of $a_1 - a_2$ versus t will yield a straight line with a slope $(\rho S)/(S_1 S_2)$, where $S = S_1 + S_2$. In determining the slope it is most convenient to find the abscissal value of time t for which $(a_1 - a_2)$ is one-half $a(0)$. From this *half value time* $t_{1/2}$ we obtain

$$\rho = \frac{0.693 S_1 S_2}{t_{1/2} S} \tag{7}$$

A critique of this method is included on pages 24–25 of the next chapter.

The integrated relations in eqs. 3 may also be used in situations where the knowledge of the system is limited. If, for instance, all of the necessary data cannot readily be obtained, the available data are compared to those predicted from the solution of the kinetic equations. We do not claim, under these circumstances, to solve for ρ, but merely to consider the two-compartment steady-state model as one of a number of alternatives. This "trying on for size" is not a particularly satisfactory procedure, but it may serve as a guide for further investigation. All too frequently this is what we are doing in current tracer practice.

Physical analogs. There are a number of physical systems whose equations are similar in form to eqs. 3. We will refer to these systems as *analog systems or analogs.* Hydrodynamic analogs are frequently described in the literature (10), and thermal analogs also exist. Suppose that a copper block is heated and placed against a cooler block in vacuo with a thin layer of material of low thermal conductivity between. In the absence of radiative loss, the flow of heat, by conduction between the blocks and its effect on their temperatures, depends upon the difference in block temperatures. The thermal conductivity of the separating layer is analogous to ρ; the temperature of a block is analogous to a; and the thermal capacity, to S.

The most interesting analogy is obtained by charging an electrical capacitor and connecting it to an uncharged one through a resistor. Here we have

$$\frac{dV_1}{dt} = \frac{1}{RC_1}(V_2 - V_1)$$

$$\frac{dV_2}{dt} = \frac{1}{RC_2}(V_1 - V_2) \tag{8}$$

In this case V_1 and V_2, the capacitor voltages, are analogous to a_1 and a_2; the conductivity $1/R$ of the resistor is analogous to ρ; and the capacitances C_1 and C_2 are analogous to S_1 and S_2. More on this subject will be found in Chapter 6, which is devoted to the physical simulation of tracer kinetic problems by analog methods.

2

Tracer Experiments in Compartmental Systems

Components of the Tracer Method

Compartments and systems. Since it is not possible to follow individual tracers by present methods, the major efforts in tracer experiments are usually devoted to observations in systems of compartments (11). The procedure begins with the introduction of labeled material into one compartment. After a suitable length of time, samples from it and from other presumed compartments are obtained and analyzed. Then the net movements between compartments are inferred from the analytical results. The analysis may include determinations of radioactivity by means of counting techniques or ion-chamber measurements. If radioactive tracers are not used, the analysis may require dye-concentration determinations or other methods of label identification. In general the analysis may also include determinations of the amount in each compartment of the substance or items being traced. Thus, if water is being traced with tritium oxide, the total amount of water in each compartment can be determined.

From these considerations it would seem that the ideal tracer experiment does not involve the solution of differential or integral equations. By a process of accounting over various time intervals in such an experiment, we simply determine transport rates by solving a set of linear simultaneous algebraic equations. The numerical values of the derivatives which would appear are known, and so the problem remains one of elementary algebra at most. This is true in principle, but, in practice, workers have preferred wherever possible to fit the experimental data points with smooth curves, using mathematical relations predicted from some type of assumed kinetic behavior of the system. To obtain these curves, we must solve the equations as differential equations and obtain specific activity relations for the compartments as functions of time.

15

Since it is instructive to discuss the tracer method from both points of view, we will consider first the simple algebraic determinations of rates in this and the following chapter. We will then follow in later chapters with kinetic analyses of some of the more important compartmental systems. This will require a discussion of the calculus of systems of simultaneous first-order linear differential equations.

We may prefer to regard some compartmental systems as open, others as closed. In the case of the body water problem we can consider the body as a one-compartment open system or the body and its environment as a two-compartment closed system. A closed system has the advantage that total S and total label are constant quantities. In general, a closed system would be any system which can be surrounded by an impermeable envelope without having its function disturbed. Of course some workers will prefer to think of a system as being closed only with respect to the substance S under consideration, but open otherwise.

In Chapter 1, the notation and definitions for a two-compartment system were indicated. These considerations will now be generalized. Figure 4 shows a general closed multicompartment system. A "compartment" may be a "physical compartment," such as all of the space within the cardiovascular system of an experimental animal. It may be a "chemical compartment," such as the bicarbonate ion in an aqueous system containing CO_2. It may also be part of a functional group in an organic molecule, such as the carboxyl carbon of acetate. In fact, any specific category in which the substance being traced is found is considered to be a compartment in this general sense. Physiologists who use the tracer method frequently use the word "pool" in place of compartment. (A compartment and a pool are considered to be synonymous.) Of course the word "compartment size" may signify a physical volume, such as the total number of liters of body water, but the basic quantity is amount, not volume. When measurements are made in grams or moles, the term "compartment size" is still generally used.

In favorable instances true compartment sizes can be obtained by the isotope dilution method and successfully compared to results from alternative methods. There are situations, of course, particularly in physiology, where the specific activity of the compartment under study does not approach a satisfactory stable value. In other instances different tracers yield conflicting results. In these adverse circumstances, estimates must be made from less than satisfactory data, and the term "space" is usually applied to the empirical values of compartment size. The "thiocyanate space" and the "Na^{24} space" in man are not equal, and neither space gives a truly satisfactory figure for the anatomical volume of the extracellular fluid.

The compartment concept is important because the contents of a true compartment are homogeneous. Because all the compartmental contents behave identically, any sample of material will be equally representative with any other sample and an outgoing sample will carry with it into any

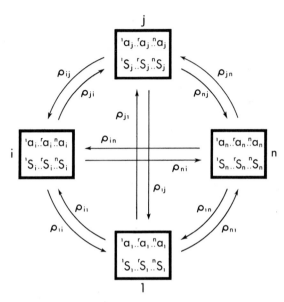

Fig. 4. Schematic diagram of a general closed multicompartment system showing the first and last compartments (1 and n) and two general intermediate compartments i and j. Rate of movement of S from compartment i to j is ρ_{ji}. If the analysis involves several species of S the amount of the rth species in compartment i is ^rS_i. For the case of radioactive tracers the specific activity of the rth tracer in compartment j is ^ra_j as shown. The specific activities may be related to total S_i and measured ^rR_i, etc. (See text.)

other compartment the same fraction of tracer. Homogeneity of compartmental contents is usually achieved by uniform mixing. Of course, if two compartments behave identically, then their contents will change with time in an identical fashion, and actual mixing will not be required. They can at once be combined into one compartment.

The utility of compartmental analysis will become less as the number of compartments in a system increases. In 1 ml of blood, it would be an absurd task to trace the movement of K^{42} from the plasma into each of the individual several billion erythrocytes. However, there are situations where many compartments which behave similarly, though perhaps not identically, can be combined or "lumped" into one equivalent compartment. Plasma proteins labeled with I^{131} can be considered as almost an infinity of compartments since the proteins are so heterogeneous. Since we cannot compartmentalize the system down to the individual proteins, we can, at best, lump the system into only a few compartments. Of course, there might be instances where systems possess no actual compartmentation. In such a system, the properties would change continuously with spatial and temporal coordinates. Except perhaps at submicroscopic levels, the smallest conceivable compartments, then, could always be further subdivided without achieving homogeneity.

Compartmental indices. In describing compartments and their contents some attention can profitably be paid to notation. Any symbol can be chosen to represent an amount of a given substance being traced. Those, for example, who have worked with phosphorus have represented it simply as P. When we represent sodium as a two-letter symbol Na, however, there is a distinct loss in economy. Recognizing that a symbol should aid the worker in keeping track of his variables, we will try in general to use letters which suggest the quantity which they represent. The desired generality will be achieved if we use the non-specific word "substance." The symbol S, without subscripts, will represent the total amount of substance in the system. This notation does have the minor drawback that we may occasionally be tracing objects, such as insect pests, rather than a substance. However, those who do tracer experiments in ecological systems will have little difficulty with the symbol S. Of course time will be represented by t.

For the identification of compartments the most convenient method is to use subscripts. In so doing, we must register objection to awkward subscripts such as S_{cell}, S_{medium}. In the present treatment we will simply number specific compartments with integers. Following conventional mathematical usage we will reserve alphabetical letters to represent compartmental indices when they are variable. Thus S_1 will be the amount of S in compartment 1, and S_i will be the amount in some general, unspecified compartment i. Usually in this procedure n is used to represent the total number of compartments. S_n, then, will represent the contents of the last in the series; S_{n-1}, the contents of the second to last, etc. Recognizing that multiple tracer experiments might be performed, we can represent several species of S through superscripts. The rth species, then, would be ^{r}S.

When S_i is constant, it is not necessary to proceed further, but in many instances compartments grow or shrink. We will then indicate that S_i varies with time by the representation $S_i(t)$, where the initial value for zero time is denoted by $S_i(0)$. The rate of change of S_i with time may be represented by the derivative dS_i/dt.

Specific activities and abundance ratios. The most important characteristic of a compartment in tracer research is the fraction of tracer which it contains. It is this characteristic which determines the chances that a molecule of S being carried from one compartment to another will be a labeled one. If the tracer is a stable isotope, such as deuterium, atoms of the heavy isotope may represent an appreciable fraction of S_i, and specification of the abundance ratio A_i may be preferred. For the case of multiple tracers $^r A_i$ would be the fraction of S in compartment i which is in the isotopic form r.

Frequently, as in the use of radioactive isotopes, it will not be necessary to measure abundance ratios but merely to determine quantities such as radioactivities. The result of such a measurement will be represented as R_i and this may be represented simply as counts per minute on some counting device of unknown but constant efficiency. Normally we will only be concerned with the total activity in compartment i and not the activity per cubic centimeter, which is usually irrelevant. When dyes are used as tracers and their concentrations read on a spectrophotometer scale, the principal interest will be in the total amount of this indicator in some portion of the system. This quantity can be computed as the product of concentration and compartment volume.

Of course R_i in radioactive experiments is not analogous to A_i. The quantity that is analogous is the "specific activity" a_i, which is the quotient R_i/S_i. It is not a very "specific" quantity at all, since it depends on the various ways in which R_i may be measured and expressed. R_i may be measured in counts per minute, but it can be microcuries or even ion pairs per second determined from electroscope readings. Frequently a_i will be represented in units of "counts per minute per millimole of S."

Equations can often be simplified if, instead of a_i, we use the ratio $a_i/a_i(0)$, where the fraction of the amount initially in the compartment is determined. If only one compartment is initially labeled, this compartment may be taken as compartment 1. However, the initial value of a is often simply called $a(0)$. Ratios of this sort involving a are usually called "relative specific activities." Although not applied in this way by Sheppard and Householder (4), this definition will be adopted here.

Transport rates. Up to this point the quantities with which we have dealt are the ones which normally should be available by direct experimental determination in the system. The derived quantities we wish to obtain from these data are the transport rates from compartment to compartment. Some experimenters prefer to designate them by various alphabetic letters such as k, M, etc. However, the use of k presents a difficulty because it has traditionally been used by chemical kineticists to denote a first-order reaction rate constant. It cannot be emphasized too strongly that tracer experiments seek an actual transport rate which measures the amount of S which really moves per unit time from compartment to compartment. There is no overwhelming reason for the particular choice of ρ for such a quantity, but if we employ R for rate we will be in conflict with the symbol for radioactivity. We have therefore preferred to use the corresponding Greek letter. No subscript is necessary in the case of pure exchange between only two compartments. Usually, however, in a multicompartment system there will be several differing rates. We will then adopt the system ρ_{ij} to represent the rate into compartment i from compartment j (Fig. 4). (Occasionally in the literature, the order of i and j is reversed.) Just as n is the maximum value of i, m will be the maximum value of j. The conventional sigma notation of mathematics will be employed to represent summation. If we wish to know the total rate at which S leaves compartment j, we must form the sum of the rates from compartment j into all of the other compartments for all values of i starting with 1 and concluding with n. If there are five other compartments, the sum will be

$$\rho_{1j} + \rho_{2j} + \rho_{3j} + \rho_{4j} + \rho_{5j}$$

In the general case the summation symbol is used, yielding

$$\sum_{i=1}^{i=n} \rho_{ij} = \rho_{1j} + \rho_{2j} + \cdots + \rho_{nj}$$

The indices beneath and above the capital sigma denote the initial and final values of i, thus indicating that i is the index on which summation is performed. Occasionally, for the sake of brevity, the summation index will be written alone under the sigma without the specification of initial and final values.

Using this convention, we can immediately proceed to the analysis of a system whose compartment sizes vary. In general, the rate of change of S, with respect to time in compartment i, can be represented by the relation

$$\frac{dS_i}{dt} = \sum_{j=1}^{j=m} \rho_{ij} - \sum_{j=1}^{j=m} \rho_{ji} \tag{1}$$

In this expression the first sum gives the total inflow rate from all other compartments and the second, the total rate of outflow.* Thus, for a general three-compartment system (see Fig. 5) the rate of change of S in compartment 2 will be

$$dS_2/dt = \rho_{21} + \rho_{23} - \rho_{12} - \rho_{32}$$

---

* Following the practice occasionally employed in physics, the symbol $\overset{\circ}{S}_i$ for dS_i/dt will sometimes be used when space is limited.

Of course the rate at which material is transported from a compartment to itself, ρ_{ii}, is meaningless and will not appear. This fact is usually expressed by placing the relation $i \neq j$ ($\neq$ means not equal to) beneath the summation sign. If this condition is not indicated, it will nevertheless be assumed

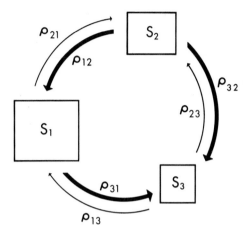

Fig. 5. A general three-compartment system showing transport from each compartment to all others.

to apply in the remainder of this book. For compartments of constant size we have the important general set of n relations, one for every value of i,

$$\sum_{\substack{j=1 \\ j \neq i}}^{j=m} \rho_{ij} - \sum_{\substack{j=1 \\ j \neq i}}^{j=m} \rho_{ji} = 0 \tag{2}$$

Communication between compartments. We can now generalize from the three-compartment system of the last section. In a general closed system of n compartments, there will be $n(n-1)$ possible values of ρ. It would simplify matters considerably if there were some topological principle whereby the permissible connections between compartments were restricted. Is there a limit to the possible connections between compartments, which are regions of ordinary space separated by boundaries? It is true that there are some limitations. We may first consider a series of points on a straight line. If three points: 1, 2, and 3, are arranged in linear order, connections between 1 and 2 and between 2 and 3 exist, but the only way to go from 1 to 3 is via 2. Such a transfer is merely the sum of transfers 1 to 2 and 2 to 3.

Of course it is not particularly meaningful to consider transport along a line, since the lateral dimensions vanish. If we proceed, then, from one dimension to two dimensions, it is possible for a region in a plane to have, at most, two neighbors with which it may communicate at one point. Figure 6 illustrates the junction between three such regions. Communication across the boundaries is possible between each of the regions and all of the others. However, if four regions instead of three meet at one point, communication can not occur between regions oppositely situated (see Fig. 6) except via the regions between. Communication through the point of intersection, of course, has no meaning since the dimensions of the point are vanishingly small.

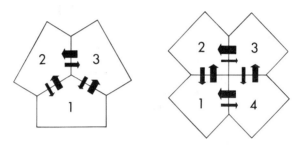

Fig. 6. Communication between regions in a plane. The three regions at the left are mutually intercommunicative. In the system at the right, communication occurs only between neighbors.

A generalization of this to three-dimensional space is not simple to represent graphically, but it is possible to place four regions of space together with intervening surfaces of separation across which communication can be achieved between any one region and all of the others. Such unrestricted communication cannot occur between a larger number. From this it may be concluded that, in a system in which physical compartments are created by a fixed system of surfaces of any sort, we need not consider that every compartment thus created must communicate with every other one. The difficulty with the topological restriction concept is that it applies to regions which communicate at one point only. Situations may occur where regions have contact at more than one point.

If different states of chemical combination are to be considered as compartments, any satisfactory general theory cannot in any event be simplified by these spatial considerations. A general multicompartment system, then, may conceivably have the maximum number of ρ's (Fig. 4). Systems with a smaller number will be referred to as constrained systems (4). Two such constrained systems are the "mammillary" system where a central compartment is surrounded by peripheral ones and the "catenary"

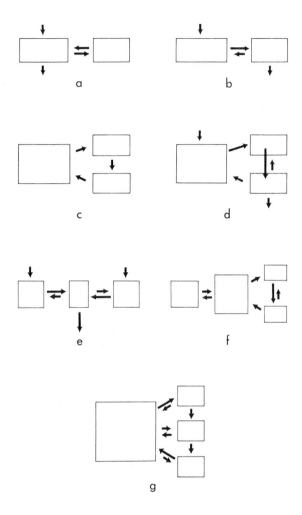

Fig. 7. Some representative schematized compartmental systems showing: (a) a two-compartment open system with turnover in one (the left) compartment and exchange with the other; (b) serial turnover plus exchange in two compartments; (c) a three-compartment closed cyclic system; (d) turnover, exchange and cyclic transport in three compartments; (e) double turnover and exchange; (f) exchange and cyclic transport in a closed three-compartment system; and (g) exchange and cyclic transport.

system where compartments are arranged in chains.* Other representative systems which may be of occasional interest are shown in Fig. 7. Although several are open systems they may be treated as closed systems containing an additional infinite compartment.

Practical Considerations in Two-Compartment Systems

Equations 1 in Chapter 1 represent the simplest multicompartment system. Determinations of R_1 and S_1 in one compartment as functions of time would suffice to establish ρ, provided the total amount of activity and of S in the entire system were known. This condition holds because only one unknown must be determined, requiring only one equation for its determination. Of course, it does not matter whether or not ρ is constant in this case, because its value at any time may be explicitly solved for, provided a_1, a_2 and either set of values—S_1 and da_1/dt or S_2 and da_2/dt—be known. This fact, of course, is true in theory.

In practice, however, an increasing precision is required as time progresses and the difference between the a's becomes smaller. Ultimately the difference becomes vanishingly small, and the required precision of the experiment becomes prohibitive. Even at the beginning of the experiment, particularly in some types of biological studies, precision may be poor. We may then wish to make multiple determinations for mutual re-enforcement. The difficulty is that often the repeated determinations will have to be performed serially and that, meanwhile, the a's may be changing. This problem is usually solved most conveniently by fitting the data to some theoretical relation. If there is adequate reason to believe that ρ is constant and that only two compartments exist, this theoretical relation is given by eq. 2 of Chapter 1.

The procedure, then, is as follows: assumptions are made concerning the theoretical behavior of the system; values of Δ_{12}, computed from experimentally determined values of radioactivity and of S, are plotted on semi-log paper; a straight line is fitted to them by statistical procedure; the slope of the line is determined and ρ is obtained by multiplying the slope by $S_1(S_2/S)$. The methods commonly used for this type of curve fitting do not weight the experimental points according to their true precision, and the slope may be biased. However, if reasonably precise values of Δ_{12} are available, a reasonably good value of ρ can be obtained. This value is usually without adequate statistical estimates of its confidence

* The term "catenated" would seem to this writer to apply to a system which was put together by plan rather than one which "just happened." The same holds for "mammillated" which, nevertheless, is considered good usage by anatomists.

limits. It is also true that this conventional approach to two-compartment kinetics must assume constant ρ and cannot in principle prove it. This is a basic limitation also in the "exponential analysis" of more general closed steady-state systems.

Three-Compartment Systems

The literature contains frequent references to tracer experiments in three-compartment systems (12, 13, 14, 15). In such a system, more than one transport rate must be determined. It now becomes necessary to consider the number of unknowns and the number of equations available for their determination. We suppose that the system is closed and in a steady state. In this case the rates between pairs of compartments may be equal and opposite, but there may also be a superimposed turnover rate in cyclic systems. For pure exchange, equations may be written, as in the two compartment case, by considering that the total rate of change of activity in a compartment is a sum of all amounts transported in minus all amounts transported out. Writing the terms on the right-hand side in a more convenient order, we have

$$S_1(da_1/dt) = \rho_{12}(a_2 - a_1) + \rho_{13}(a_3 - a_1)$$
$$S_2(da_2/dt) = \rho_{12}(a_1 - a_2) + \rho_{23}(a_3 - a_2)$$
$$S_3(da_3/dt) = \rho_{13}(a_1 - a_3) + \rho_{23}(a_2 - a_3) \tag{3}$$

Since the processes are those of simple exchange, ρ_{ij} is equal to ρ_{ji}.

It would appear at first that this set of three equations could be solved for the three ρ values, but an important relation exists among them which makes solution impossible. If the first two equations are added together, the right-hand side becomes the negative of the right-hand side of the third. This condition means that the third equation is incompatible with the others unless the expression

$$S_1 \frac{da_1}{dt} + S_2 \frac{da_2}{dt} + S_3 \frac{da_3}{dt} = 0 \tag{4}$$

holds among the left-hand quantities. If this relation is true, then the third equation is derivable from the other two and adds nothing new to the relations among the ρ's. Actually the relation *is* true in a closed system and expresses the principle that the total radioactivity in the system is constant. Although ratios of two of the ρ's to a third may be obtained from two of the equations, three ρ values are unobtainable. We can therefore conclude that no single tracer experiment using only one label can enable an investigator to determine the three possible exchange rates in a three-compartment system. If, however, one of the rates is known from other data and, in particular, if that rate is zero, the other two can then be determined. Such a system, with one or more zero rates, is a constrained system. Constraints can in some instances make a system soluble which is otherwise insoluble for lack of data.

What can be done if the system is not constrained in any way and if none of the rates is known? Two possibilities exist by which additional equations can be obtained. One is to determine a new set of a's and da/dt's from a second experiment. If the same initial distribution of radioactivity is used and the same ρ's are expected to hold, then nothing new will be added. However, if the experiment is conducted under identical conditions so far as the ρ's and S's are concerned, but with a different initial distribution of label it will yield a new set of a's. The other alternative is to use two different tracers with different initial distributions and establish among the a's for the second tracer a second set of relations of the same form as eqs. 3. Whichever alternative is used the additional equations which result will then provide the necessary data for a solution for all of the required ρ's.

Incomplete Mixing

The assumption of homogeneity in compartments is one of the cornerstones of the tracer method, but all too frequently it is not well founded. In physiological experiments particularly, this assumption may present a serious problem. A possible illustration is the kinetics of exchange of fluid between the circulation of man or experimental animals and the extracellular fluid. If the substance under study is one which exchanges relatively slowly compared with rates of circulatory mixing, the intravascular compartment may be considered as reasonably well mixed. The difficulty is in the assumption of uniform mixing outside the circulation. Samples from certain extravascular compartments are relatively easy to obtain in animals by techniques such as lymphatic cannulation. However, other compartments, such as the tenuous extracellular fluid layer bathing striated muscle fibers are essentially inaccessible.

For this reason and others it is of interest to obtain a qualitative picture of the sort of errors that can occur as a result of failure of the mixing postulate to a greater or lesser degree. Figure 8 illustrates one way in which inhomogeneity in a second compartment can affect the result in a steady-state pseudo two-compartment system. In this case the inhomogeneity is expressed by dividing compartment 2 into a pair of subsidiary compartments 2' and 2". These subsidiary compartments together with the original compartment form a mammillary system. Initially the specific activity changes in the initial compartment are not altered by the inhomogeneity.

If the two subsidiary compartments are taken as one in the determination of a mean specific activity a_2, it will still increase linearly in proportion to the sum of ρ' and ρ''. These are the two different rates at which tracer is brought in, and the sum will be considered as a single overall rate ρ. As the contents of compartments $2'$ and $2''$ begin to contain more label,

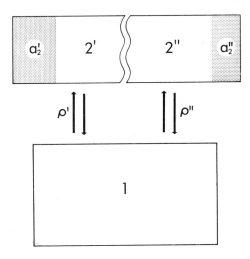

Fig. 8. A pseudo two-compartment system. Compartment 2 is represented by two subcompartments with specific activities a_2' and a_2'' and exchange rates ρ' and ρ''.

backflow into compartment 1 will occur. If the contents of the two subsidiary compartments were homogeneous with a uniform specific activity a_2, the backflowing material would carry $a_2\rho$ units of radioactivity per unit of time back into compartment 1. Now, however, some of the returning material has specific activity a_2' and some, a_2'', the overall figure being a mean weighted in proportion to the separate exchange rates ρ' and ρ''. The effect of inhomogeneity will depend then on how great the difference is between $a_2\rho$ and $a_2'\rho' + a_2''\rho''$.

It will require a considerable difference in the properties of the two compartments before much effect will be seen in this case. We may illustrate by supposing, using approximate arithmetical estimates, that the two subsidiary compartments are of equal size, each containing one mole of S. Let ρ' be 0.02 mole per hour and ρ'' be 0.01 mole per hour, so that ρ is 0.03 mole per hour. If initially a_1 is 1 microcurie per mole, then during the first 30 minutes 0.015 microcurie will appear in compartment 2 (taking only tracer which makes its first appearance). The average specific activity will be 0.015/2 = 0.0075 microcurie per mole. Individual values will be $a_2' = 0.01$, $a_2'' = 0.005$. The return flow of tracer will be $2 \times 10^{-4} + 5 \times 10^{-5} = 2.5 \times 10^{-4}$ microcurie per hour. If the two compartments were combined and the combination uniformly mixed, the rate would be $a_2 = 0.0075 \times 0.03 = 2.25 \times 10^{-4}$ microcurie per hour. In one case the net rate of decline of activity in compartment 1 is $0.03 - 0.00025 = 0.029750$ microcurie per hour, in the other case 0.029775 microcurie per hour.

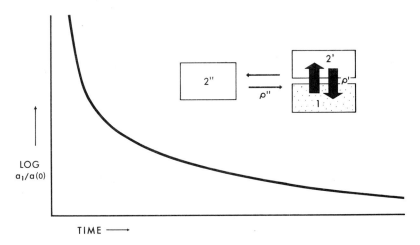

Fig. 9. Specific activity in the central compartment of a mammillary system with widely different peripheral exchange rates. One peripheral compartment (large arrows) rapidly equilibrates followed by later slow exchange with the second. Stippling in compartment 1 indicates initial label.

At this time the deviation in behavior in compartment 1 would scarcely be observable. Slight evidence could be found by observations made in this compartment that there was an inhomogeneous situation in the presumed single second compartment. Sheppard and Householder (4) have indicated the behavior of such a system in more detail. In a mammillary system, unless the behavior of two peripheral compartments differs by a considerable amount, the results can be expressed surprisingly well in terms of a single lumped compartment, particularly during early time following labeling.

Of course a highly inhomogeneous second compartment can soon influence the results in the first. Suppose that in the previous analysis ρ' is twenty times greater than ρ''. The two compartments with the rapid exchange will equilibrate almost completely before much exchange of label has occurred with the slow compartment. The two fast compartments will interact with the slow one almost as though they were one uniformly mixed system. Thus in the initially labeled compartment there will be a rapid initial drop superimposed on a slowly declining base line. If a semi-log plot is constructed (Fig. 9), it will show a considerable convexity toward the origin. In general, inhomogeneity in a system will usually exhibit itself as a deviation of a semi-log plot from expected linearity.

In a mammillary system there is a tendency for inhomogeneity to average out to some extent. If one of the samples entering a compartment originates in some area with above-average specific activity, other samples

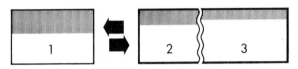

Fig. 10. Exchange in a pseudo two-compartment catenary system. Higher specific activity (indicated by shading) in compartment 2 causes label returning to compartment 1 to be higher than it would be if 2 and 3 formed a single homogeneous compartment. The result is a decreased rate of loss of activity from compartment 1.

coming from below-average regions may partly compensate for the inhomogeneity. In a catenary type of inhomogeneity, however (Fig. 10), the rate of decline of specific activity in the initially labeled compartment (compartment 1 in the figure) is influenced by the rate at which the immediate neighbor (compartment 2) returns label to it. This is higher if its specific activity is higher as it will be in the absence of rapid equilibration between 2 and its more remote subsidiary 3. This sluggishness between 2 and 3 can introduce an overall rate-controlling effect on the movement of label out of compartment 1. This situation will have a systematic biasing effect on the results.

3

Theory of Tracer Experiments in Multicompartment Systems

General Principles

In considering an analysis of the completely unrestricted multicompart-ment system, those who merely use tracers as a tool may find that the results appear unnecessarily academic. Nevertheless, to others who wish to solidify their intuition and who are interested in logical foundations, the considerations may be highly instructive. The generalizing of some of the material presented in Chapters 1 and 2 requires a slightly more advanced mathematical approach. Solutions of systems of linear equations may be expressed in determinant form by algebraic methods described in many standard reference textbooks (16). As before we will consider closed systems since most open systems can be converted to closed systems by adding compartments.

In a general closed system of n compartments (Fig. 4) we may note that, excluding communication with itself, each compartment can communicate with $n - 1$ others. Multiplying the number of interactions for one compartment by the number n of compartments yields for the total number of interactions $n(n - 1)$. If we are to use labels to determine the kinetic behavior of such a system, it will be necessary to relate the individual transport rates from compartment to compartment to the movement of labels between the compartments. Such movement is restricted, by our assumed methodology, to analysis of the compartmental contents, a procedure which requires access to the compartments being analyzed. If we are to obtain all of the rates uniquely, then the results of our experiments must yield as many equations as there are unknowns. Lacking the necessary information we must be satisfied merely with the determination of some of the rates in terms of others which might in some way be known or inferred from further data.

If maximum generality is to be retained in the analysis, we should consider separately all species of S involved, without selecting one species as having special importance over another. Thus, for example, if we are performing experiments with a mass spectrometer using various identifiable stable isotopes, we will consider the relative abundances of all of the various species. An analysis has already been made of the theory of multicompartment systems in terms of abundance ratios as measured with the mass spectrometer (11). Actually, however, the typical tracer experiment, as we know it, usually involves the tracing of some predominant substance in the presence of small amounts of one or more isotopic species of the same material. These minor species then serve as labels.

We will first consider the case of one label only, where the amount of substance S in the ith compartment is S_i, its total activity is R_i, and the specific activity R_i/S_i in compartment i is a_i. The transport rate into compartment i from compartment j is ρ_{ij}. The equations can then be written out, starting in compartment 1, as follows:

$$dR_1/dt = \rho_{12}a_2 + \rho_{13}a_3 + \cdots \rho_{1n}a_n - \rho_{21}a_1 - \rho_{31}a_1 \cdots - \rho_{n1}a_1$$

$$dR_2/dt = \rho_{21}a_1 + \rho_{23}a_3 + \cdots \rho_{2n}a_n - \rho_{12}a_2 - \rho_{32}a_2 \cdots - \rho_{n2}a_2$$

$$\cdots \cdots \cdots \cdots \cdots \cdots \cdots \cdots \cdots \cdots \cdots \cdots \cdots \cdots \cdots \cdots \cdots \cdots$$

$$dR_n/dt = \rho_{n1}a_1 + \rho_{n2}a_2 + \cdots \rho_{n,n-1}a_{n-1} - \rho_{1n}a_n - \rho_{2n}a_n \cdots$$
$$- \rho_{n-1,n}a_n \quad (1)$$

Since transport from a compartment into itself yields no information in a tracer experiment, all terms with identical subscripts are omitted.

It is readily apparent that this method of representing a general system is very cumbersome. The actual meaning can be expressed by using the simpler summation notation which only requires writing a single equation for compartment i. It should be recognized that i will have all possible values from 1 to n, where n represents the number of compartments in any given case. Thus we write

$$\frac{dR_i}{dt} = \sum_{\substack{j=1 \\ j \neq i}}^{j=n} \rho_{ij}a_j - a_i \sum_{\substack{j=1 \\ j \neq i}}^{j=n} \rho_{ji}$$

or more concisely yet,

$$\frac{dR_i}{dt} = \sum_{\substack{j \\ j \neq i}} [\rho_{ij}a_j - a_i\rho_{ji}] \quad (2)$$

Since there are as many j as i the upper limit for the sum will be n. The expression $j \neq i$ means that all terms with identical subscripts do not appear. This more concise method of representation will be used in the material which follows, but occasions will arise when the relations will be written out in expanded form. This will be for the purpose of verifying some particular relation or appreciating the full implication of certain expressions.

Upon examining the n equations of eq. 2 we see clearly that there are not enough of them to determine the $n(n - 1)$ ρ's. It is a familiar mathematical principle (16) that each equation of a system imposes a new relation among the unknowns by setting a restriction on the way in which they may vary independently. If there are as many restrictions as there are unknowns, then none of the unknowns may vary, because all are uniquely determined. Given any set of equations, however, we can make up as many equations as we desire by merely expressing in different ways the same relation or the same restriction of the variables. None of these additional redundant equations adds anything to the restrictions of the system.

These further equations, then, cannot be accepted as valid in uniquely determining the solutions of the system under consideration. In the present instance we can arbitrarily select any one of the equations in (2) as redundant. Suppose that all the equations but the first are added together and the resulting combination multiplied through by a minus sign. The left-hand side becomes identical with that of the equation for compartment 1. (In any closed system the total activity is constant. Thus the negative of the sum of dR_i/dt in all compartments but the first equals dR_1/dt.) Similarly adding up the terms on the right-hand side of all equations but the first and reversing the signs yields the right-hand side of the first. Therefore, the first equation is merely an alternative expression of the others. The same conclusion can be drawn for any particular one of the set which we may select. Thus there are only $n - 1$ actual relations which can be used to solve for the $n(n - 1)$ unknowns in this system.

One further set of equations is obtained from the relations between rates of change of S in each compartment. Thus

$$\frac{dS_i}{dt} = \sum_{\substack{j \\ j \neq i}} (\rho_{ij} - \rho_{ji}) \tag{3}$$

But, because $\sum_i dS_i/dt = 0$ in a closed system, only $n - 1$ of these relations are independent. At this point $n - 2$ more sets of $n - 1$ relations are required.

There are two ways in which the necessary information can be obtained. We may repeat the same experiment under identical conditions but with a different distribution of label so that the values of the a's will be different.

Practically speaking, the difference should be as great as possible. Otherwise, if the a's are the same as in a previous experiment, the equations will provide no new information. The alternative is to perform experiments with multiple tracers, with each tracer having a different initial distribution in the system. Of course the use of multiple tracers is equivalent to multiple experiments performed simultaneously. If neither of the alternatives is acceptable, and this will most frequently be the case, then we are compelled to accept less information and solve for only certain of the ρ's in terms of others which might be known from other data. In general, then, systems having minimal numbers of compartments must be combined with maximal ingenuity in using ancillary information if the tracer method is to succeed.

Assuming, for purposes of theoretical discussion, that multiple tracers or multiple experiments can be obtained, we can identify them from one another by the use of a superscript r. The analysis can also apply to a mixed situation with a few tracers and more than 1 but less than n experiments. The specific activity in compartment i during experiment r for example would be $^{r}a_i$, etc. The final set of equations, then, would be

$$\frac{d(^{r}R_i)}{dt} = \sum_{\substack{j \\ j \neq i}} \rho_{ij}{}^{r}a_j - {}^{r}a_i \sum_{\substack{j \\ j \neq i}} \rho_{ji} \tag{4}$$

for $n - 1$ values of r, and also

$$\frac{dS_i}{dt} = \sum_{\substack{j \\ j \neq i}} (\rho_{ij} - \rho_{ji}) \tag{5}$$

If determinations of the rates of change of the S's are lacking, the last set can be replaced with a new experiment in which data for one more value of r could be obtained. Thus, although total S can be used as a "tracer," it is not essential if a substitute tracer can be found. Only one set of relations among the S's is available as a tracer since, if the experiments are to be identical, the same variations in the S's will always occur on each repetition.

In writing the equations it is advantageous to preserve some order. We will first consider compartment $1(i = 1)$, and write in order, one below the other, the relations for the first experiment, the second, and so on ($r = 1, 2,$ etc.). Then a similar series will be written for compartment 2, and the progression is continued on up to compartment $n - 1$. In this way the unknown ρ's will be arranged in a square array of rows and columns. In this columnar arrangement, the ρ's in each row will occur in groups. In each group the first subscript i of the separate terms will be identical while the value of the second subscript j increases in order from 1 to n. The groups in each row, in turn, will be arranged in increasing order of the first subscript i from 1 to $n - 1$. In each group a term will be missing corresponding to $i = j$, but symmetry will be preserved by inserting in this missing position the term containing ρ_{n1}, or ρ_{n2}, etc., as the case may be. Here is an illustration of the arrangement for $n = 3$:

$$- {}^1a_1\rho_{31} + {}^1a_2\rho_{12} + {}^1a_3\rho_{13} - {}^1a_1\rho_{21} + \quad 0 \quad + \quad 0 \quad = \frac{d({}^1R_1)}{dt}$$

$$- {}^2a_1\rho_{31} + {}^2a_2\rho_{12} + {}^2a_3\rho_{13} - {}^2a_1\rho_{21} + \quad 0 \quad + \quad 0 \quad = \frac{d({}^2R_1)}{dt}$$

$$- {}^3a_1\rho_{31} + {}^3a_2\rho_{12} + {}^3a_3\rho_{13} - {}^3a_1\rho_{21} + \quad 0 \quad + \quad 0 \quad = \frac{d({}^3R_1)}{dt}$$

$$0 \quad - {}^1a_2\rho_{12} + \quad 0 \quad + {}^1a_1\rho_{21} - {}^1a_2\rho_{32} + {}^1a_3\rho_{23} = \frac{d({}^1R_2)}{dt}$$

$$0 \quad - {}^2a_2\rho_{12} + \quad 0 \quad + {}^2a_1\rho_{21} - {}^2a_2\rho_{32} + {}^2a_3\rho_{23} = \frac{d({}^2R_2)}{dt}$$

$$0 \quad - {}^3a_2\rho_{12} + \quad 0 \quad + {}^3a_1\rho_{21} - {}^3a_2\rho_{32} + {}^3a_3\rho_{23} = \frac{d({}^3R_2)}{dt} \quad (6)$$

In general, using the running index system and the summation notation, we may write the expression in the following abbreviated form

$${}^ra_1\rho_{i1} + \cdots {}^ra_{i-1}\rho_{i,i-1} - {}^ra_i\rho_{ni} + {}^ra_{i+1}\rho_{i,i+1}$$

$$+ \cdots {}^ra_n\rho_{in} - \sum_{\substack{j=1 \\ j \neq i}}^{j=n-1} {}^ra_i\rho_{ji} = \frac{d({}^rR_i)}{dt} \quad (7)$$

Solutions of equations of this type are obtained by familiar algebraic methods through the use of determinants. Readers who are acquainted with these methods will proceed directly to the conclusions without reference to the following theorems presented without proof.

Theorem I.   Given a system of n simultaneous equations as follows:

$$a_1x + b_1y + c_1z + \cdots = d_1$$
$$a_2x + b_2y + c_2z + \cdots = d_2$$
$$\cdots\cdots\cdots\cdots\cdots\cdots$$
$$a_nx + b_ny + c_nz + \cdots = d_n \quad (8)$$

n unique solutions may then be expressed as follows:

$$x = \frac{\Delta_x}{\Delta}, \qquad y = \frac{\Delta_y}{\Delta}, \qquad z = \frac{\Delta_z}{\Delta}, \text{ etc.} \quad (9)$$

Here Δ is a determinant of "rank n" represented by the square array of n rows and n columns containing all the coefficients on the left side of the system as follows:

$$\Delta = \begin{vmatrix} a_1 & b_1 & c_1 & \cdots \\ a_2 & b_2 & c_2 & \cdots \\ a_3 & b_3 & c_3 & \cdots \\ \cdots & \cdots & \cdots & \\ a_n & b_n & c_n & \cdots \end{vmatrix} \tag{10}$$

The theorem holds only when Δ is non-zero. Δ_x, Δ_y, Δ_z, etc. are similar determinants obtained by replacing the column of coefficients of x, y, z, etc., respectively, by the column of constants d, etc., on the right side of the equations.

In obtaining the numerical value of a determinant, we find that further theorems are useful.

Theorem 2. Provided that a row or column is at the same time retained, it can also be multiplied by a constant and added to any other row or column without altering the value of a determinant. The constant may, for example, be "-1."

Theorem 3. A determinant which is represented by a series of n diagonally located subdeterminants is equal to the product of the subdeterminants. The theorem also applied to "subdeterminants" which are of rank 1, i.e., single elements. Thus

$$\begin{vmatrix} a & b & 0 & 0 \\ c & d & 0 & 0 \\ 0 & 0 & e & f \\ 0 & 0 & g & h \end{vmatrix} = \begin{vmatrix} a & b \\ c & d \end{vmatrix} \begin{vmatrix} e & f \\ g & h \end{vmatrix}$$

$$\begin{vmatrix} a & 0 & 0 \\ 0 & b & 0 \\ 0 & 0 & c \end{vmatrix} = abc \tag{11}$$

Theorem 4. In the case of a determinant of the type in theorem 3, inserting any elements into the zero positions on one side of the diagonal will not alter its value:

$$\begin{vmatrix} a & b & 0 & 0 \\ c & d & 0 & 0 \\ p & x & a & b \\ q & y & c & d \end{vmatrix} = \begin{vmatrix} a & b & 0 & 0 \\ c & d & 0 & 0 \\ 0 & 0 & a & b \\ 0 & 0 & c & d \end{vmatrix} = \begin{vmatrix} a & b \\ c & d \end{vmatrix}^2$$

$$\begin{vmatrix} a & 0 & 0 \\ p & b & 0 \\ q & r & c \end{vmatrix} = abc \tag{12}$$

The proof of theorems 3 and 4 may be obtained by the use of the Laplace expansion theorem (16).

Theorem 5.   The effect of multiplying all members of a row or column of a determinant by -1 is to change the sign of the value of the determinant.

Theorem 6.   If a can be factored out of all of the elements of a row or column of a determinant, it may be factored out as a common multiplier of the determinant:

$$\begin{vmatrix} ax & y \\ az & w \end{vmatrix} = a \begin{vmatrix} x & y \\ z & w \end{vmatrix}$$

Theorem 7.   If any pair of rows or columns contains identical elements, the value of the determinant is zero.

Theorem 8.   A determinant is a polynomial which is equal to the sum of all products obtained by multiplying each element of a column by its minor with appropriate sign. The minor is defined as the subdeterminant obtained by eliminating the row and column to which the element belongs in each case. The process so defined is called "expansion on minors" and can also be performed with elements of a row instead of a column. Signs of the elements depend upon their position in the determinant and, as a result, alternate as in the following expansion:

$$\begin{vmatrix} a & b & c \\ d & e & f \\ g & h & i \end{vmatrix} = a \begin{vmatrix} e & f \\ h & i \end{vmatrix} - d \begin{vmatrix} b & c \\ h & i \end{vmatrix} + g \begin{vmatrix} b & c \\ e & f \end{vmatrix}$$

This theorem permits the numerical evaluation of a determinant.

Solution of the Equations

The case where $n = 3$ will illustrate the principles involved in obtaining solutions of eq. 7. Its generalization to other values of n is obvious. Thus,

$$\Delta = \begin{vmatrix} -^1a_1 & ^1a_2 & ^1a_3 & -^1a_1 & 0 & 0 \\ -^2a_1 & ^2a_2 & ^2a_3 & -^2a_1 & 0 & 0 \\ -^3a_1 & ^3a_2 & ^3a_3 & -^3a_1 & 0 & 0 \\ 0 & -^1a_2 & 0 & ^1a_1 & -^1a_2 & ^1a_3 \\ 0 & -^2a_2 & 0 & ^2a_1 & -^2a_2 & ^2a_3 \\ 0 & -^3a_2 & 0 & ^3a_1 & -^3a_2 & ^3a_3 \end{vmatrix}$$

To simplify, change the signs of the members both in column 1 and in column 5, thereby not changing the sign of Δ. Replace column 4 by the sum of columns 1 and 4. Then replace column 2 by the sum of columns 5 and 2. The result will be the following expression:

$$
\Delta = \begin{vmatrix}
{}^1a_1 & {}^1a_2 & {}^1a_3 & 0 & 0 & 0 \\
{}^2a_1 & {}^2a_2 & {}^2a_3 & 0 & 0 & 0 \\
{}^3a_1 & {}^3a_2 & {}^3a_3 & 0 & 0 & 0 \\
0 & 0 & 0 & {}^1a_1 & {}^1a_2 & {}^1a_3 \\
0 & 0 & 0 & {}^2a_1 & {}^2a_2 & {}^2a_3 \\
0 & 0 & 0 & {}^3a_1 & {}^3a_2 & {}^3a_3
\end{vmatrix}
$$

$$
= \begin{vmatrix}
{}^1a_1 & {}^1a_2 & {}^1a_3 \\
{}^2a_1 & {}^2a_2 & {}^2a_3 \\
{}^3a_1 & {}^3a_2 & {}^3a_3
\end{vmatrix}^2
\tag{13}
$$

We can now express the solution of the equations, for example, for ρ_{12} as

$$
\rho_{12} = \frac{1}{\Delta}\begin{vmatrix}
-{}^1a_1 & d({}^1R_1)/dt & {}^1a_3 & -{}^1a_1 & 0 & 0 \\
-{}^2a_1 & d({}^2R_1)/dt & {}^2a_3 & -{}^2a_1 & 0 & 0 \\
-{}^3a_1 & d({}^3R_1)/dt & {}^3a_3 & -{}^3a_1 & 0 & 0 \\
0 & d({}^1R_2)/dt & 0 & {}^1a_1 & -{}^1a_2 & {}^1a_3 \\
0 & d({}^2R_2)/dt & 0 & {}^2a_1 & -{}^2a_2 & {}^2a_3 \\
0 & d({}^3R_2)/dt & 0 & {}^3a_1 & -{}^3a_2 & {}^3a_3
\end{vmatrix}
$$

$$
= \frac{1}{\Delta}\begin{vmatrix}
-{}^1a_1 & d({}^1R_1)/dt & {}^1a_3 & 0 & 0 & 0 \\
-{}^2a_1 & d({}^2R_1)/dt & {}^2a_3 & 0 & 0 & 0 \\
-{}^3a_1 & d({}^3R_1)/dt & {}^3a_3 & 0 & 0 & 0 \\
0 & d({}^1R_2)/dt & 0 & {}^1a_1 & -{}^1a_2 & {}^1a_3 \\
0 & d({}^2R_2)/dt & 0 & {}^2a_1 & -{}^2a_2 & {}^2a_3 \\
0 & d({}^3R_2)/dt & 0 & {}^3a_1 & -{}^3a_2 & {}^3a_3
\end{vmatrix}
$$

$$
= \frac{1}{\Delta}\begin{vmatrix}
{}^1a_1 & d({}^1R_1)/dt & {}^1a_3 \\
{}^2a_1 & d({}^2R_1)/dt & {}^2a_3 \\
{}^3a_1 & d({}^3R_1)/dt & {}^3a_3
\end{vmatrix}
\times
\begin{vmatrix}
{}^1a_1 & {}^1a_2 & {}^1a_3 \\
{}^2a_1 & {}^2a_2 & {}^2a_3 \\
{}^3a_1 & {}^3a_2 & {}^3a_3
\end{vmatrix}
\tag{14}
$$

Finally, we can divide out the common determinantal factors in the numerator and the denominator and obtain

$$
\rho_{12} = \frac{\begin{vmatrix} {}^1a_1 & d({}^1R_1)/dt & {}^1a_3 \\ {}^2a_1 & d({}^2R_1)/dt & {}^2a_3 \\ {}^3a_1 & d({}^3R_1)/dt & {}^3a_3 \end{vmatrix}}{\begin{vmatrix} {}^1a_1 & {}^1a_2 & {}^1a_3 \\ {}^2a_1 & {}^2a_2 & {}^2a_3 \\ {}^3a_1 & {}^3a_2 & {}^3a_3 \end{vmatrix}} \tag{15}
$$

We note that the numerator determinant is obtained by inserting in column 2 (second subscript of ρ) the values of dR/dt for compartment 1 (first subscript).

This method of solution is satisfactory except when the first subscript is n, which, in this case, is 3. If the unknown (here ρ_{31} or ρ_{32}) is to appear in the proper position for a solution to be obtained, we must not omit the redundant last set of equations for compartment n, but we may omit any other set instead. A simpler device here is to solve the present set for the case where the subscripts are reversed and insert corresponding values in the determinants. This action is valid because there is complete symmetry in the subscripts throughout. For example, to solve for ρ_{31} we first obtain the numerator for ρ_{13} by inserting the dR_1/dt's in column 3. Thus

$$
\rho_{13} = \frac{\begin{vmatrix} {}^1a_1 & {}^1a_2 & d({}^1R_1)/dt \\ {}^2a_1 & {}^2a_2 & d({}^2R_1)/dt \\ {}^3a_1 & {}^3a_2 & d({}^3R_1)/dt \end{vmatrix}}{\begin{vmatrix} {}^1a_1 & {}^1a_2 & {}^1a_3 \\ {}^2a_1 & {}^2a_2 & {}^2a_3 \\ {}^3a_1 & {}^3a_2 & {}^3a_3 \end{vmatrix}} \tag{16}
$$

When the indices are reversed, we insert the dR/dt's for compartment 3 in column 1. Therefore

$$
\rho_{31} = \frac{\begin{vmatrix} d({}^1R_3)/dt & {}^1a_2 & {}^1a_3 \\ d({}^2R_3)/dt & {}^2a_2 & {}^2a_3 \\ d({}^3R_3)/dt & {}^3a_2 & {}^3a_3 \end{vmatrix}}{\begin{vmatrix} {}^1a_1 & {}^1a_2 & {}^1a_3 \\ {}^2a_1 & {}^2a_2 & {}^2a_3 \\ {}^3a_1 & {}^3a_2 & {}^3a_3 \end{vmatrix}} \tag{17}
$$

This rather cumbersome method has been chosen for deriving results in order that a knowledge of matrix theory will not be required. Actually, the latter approach is not difficult to master, and its use can be found in the literature on tracer theory. A brief introduction will be found in Chapter 11.

In general the solution is

$$
\rho_{jk} = \frac{\begin{vmatrix} a_1 \cdots {}^1a_{k-1} & d({}^1R_j)/dt & {}^1a_{k+1} \cdots {}^1a_n \\ \cdots\cdots\cdots\cdots\cdots\cdots\cdots\cdots \\ \cdots\cdots\cdots\cdots\cdots\cdots\cdots\cdots \\ {}^na_1 \cdots {}^na_{k-1} & d({}^nR_j)/dt & {}^na_{k+1} \quad {}^na_n \end{vmatrix}}{\begin{vmatrix} {}^1a_1 \cdots\cdots\cdots\cdots\cdots\cdots\cdots {}^1a_n \\ \cdots\cdots\cdots\cdots\cdots\cdots\cdots\cdots \\ {}^na_1 \cdots\cdots\cdots\cdots\cdots\cdots {}^na_n \end{vmatrix}}
\tag{18}
$$

In each case the numerator is obtained by replacing the kth row by the dR_j/dt's.

To show, for example, that in the single experiment case the nth isotope need not be determined, we must consider a relation among the specific activities in a compartment, say, the jth. Each specific activity of the rth isotope ${}^ra_j$ is related to its abundance ratio ${}^rA_j$. Thus

$$
{}^rA_j = {}^rC \, {}^ra_j
\tag{19}
$$

where the constant ${}^rC$ reflects the efficiency of the G. M. counter or some other detecting instrument and also reflects the units employed in determining the specific activity. This factor will always be the same in all compartments but may have various values for the various isotopes. A relation similar to eq. 19 also holds both for the actual amount of label and for its rate of change with time, thus

$$
\frac{d({}^rS_j)}{dt} = {}^rC \frac{d({}^rR_j)}{dt}.
\tag{20}
$$

If all of the substance S is represented by the various species ${}^rS$,

then $\sum_r {}^rS_j = S_i$ and $\sum_r \frac{d({}^rS_j)}{dt} = \sum_r {}^rC \frac{d({}^rR_j)}{dt} = \frac{dS_j}{dt}.$ (21)

Similarly, the sum of all the fractional amounts will be unity, provided all are included, so

$$
\sum_r {}^rA_j = \sum_r {}^rC \, {}^ra_j = 1
\tag{22}
$$

In the case of radioactive isotopes, of course, the C factors for all but the last species may be very small; the last would be represented as the non-radioactive or the "unlabeled" species.

We may apply these considerations to eq. 18 by multiplying each row in turn by ^1C, ^2C, etc., and adding it to the last. The final row of the denominator determinant will have the form

$$\sum_r {}^rC\,{}^ra_1,\ \sum_r {}^rC\,{}^ra_2 \cdots \sum_r {}^rC\,{}^ra_n \tag{23}$$

which, under the conditions of relation 22 becomes a series of 1's. Similarly in the last row of the numerator there will be a series of 1's everywhere except in the kth column which, because of relation 21, will contain $d(S_j/dt)$, i.e., the total rate of change of S in compartment j. The final relation, then, is

$$\rho_{jk} = \frac{\begin{vmatrix} {}^1a_1 \cdots {}^1a_{k-1} & d({}^1R_j)/dt & {}^1a_{k+1} \cdots {}^1a_n \\ \cdots\cdots\cdots\cdots\cdots\cdots \\ 1 \qquad 1 \qquad d(S_j/dt) \qquad 1 \qquad 1 \end{vmatrix}}{\begin{vmatrix} {}^1a_1 \cdots\cdots\cdots\cdots\cdots {}^1a_n \\ \cdots\cdots\cdots\cdots\cdots\cdots \\ 1 \cdots\cdots\cdots\cdots\cdots\cdots 1 \end{vmatrix}} \tag{24}$$

From this relation we conclude that, if an experiment with radioactive tracers is performed in an unconstrained multicompartment system of n compartments, determination of all the unknown transport rates between compartments can be achieved if the following data are available:

a. Determination of specific activities of $n - 1$ tracers in all compartments.

b. Determination of rates of change of activities of $n - 1$ tracers in the receiving compartment.

c. Determination of rate of change of total S in the receiving compartment.

Entry into all compartments will be necessary unless the required values of the quantities are inferred from other data. An analysis for the situation of non-radioactive tracers and for differential changes in R and S, rather than rates of change, may also be constructed along similar lines (11). Since the equations relate the ρ's uniquely to the data, it is clear that no solutions can be obtained if any of the data are missing. Thus the criteria are necessary as well as sufficient.

Experiments Late in Time

As tracers are allowed to equilibrate in a system, the various species of S will mix with one another, and ultimately the fraction of each in all compartments will be equal. In experiments with radioactive tracers all specific activities will tend toward equality throughout the system. Physically the entropy will tend to increase. Information content will decrease, and the system will proceed to a less ordered configuration. Once the system has thus "run down," it will no longer be possible to obtain the ρ values. We appreciate this state of affairs when we inspect the denominator of (18). (If all of the a's are alike in any two compartments, then, according to theorem 7, two columns contain identical elements so the determinant vanishes.) Since Δ now is zero, theorem 1 fails. When two compartments have equal specific activities, tracing between them is obviously no longer possible.*

The determinant also vanishes if the a's for any two species are alike in all compartments, or, according to theorem 6, if any two species are everywhere proportional to one another. Here Δ contains identical rows. Two tracers of equal or proportional specific activity are no better than one tracer.

Of course, from a practical point of view, Δ will never actually vanish. We thus interpret the condition as occurring effectively when the value of Δ becomes small enough that significant error is produced in the ρ's by instrumental uncertainty, truncation errors or other sources of "noise." Whether large or small the determinant serves as a measure of tracer efficiency. Although Δ may effectively vanish it may be possible in occasional instances to obtain values of a smaller number of ρ's in a reduced system.

Initial Rates of Change

Solutions of multicompartment systems are often simplified if observations are made early in time. In such situations the values in initially unlabeled compartments may often be low enough to permit the assumption of zero specific activity. For example, let us suppose that two compartments of a multicompartment system are of particular interest and are readily accessible. The initial labeling of one compartment establishes a known transport rate between it and the other compartment, because no other compartments can contribute tracer. The initial rate of appearance of label in the unlabeled compartment yields the transport rate from the labeled one. If, by chemical determination, a steady state is found to

---

* If a system has run down so that column elements in Δ approach one another numerically it may be possible by further injection of label to revitalize it. In this case the object will be to produce column differences by appropriate redistribution of label. This approach has been used in certain cases by Wrenshall and his associates in their "multiple injection technique" (17).

exist, the reverse rate is also known. Of course the initial rate of disappearance of label from the labeled compartment only yields the sum of all outward transport rates.

Multiple experiments can be represented by initially labeling various compartments in turn and observing the changes early in time in each case. In the case of multiple tracers or multiple experiments the rate ρ_{jk} can be determined by measuring the initial rate of appearance in compartment j of the label initially placed in compartment k. In general early observations made while the denominator determinant has its greatest numerical value yield information which is more precise than that later obtained.

Inferences in Compartments Which Cannot be Entered

Much of the present art of the tracer method lies in the correct inferences to be drawn concerning the contents of compartments which cannot be entered. In some compartments only determinations of a are required, and, as with any "intensive" quantity, a small sample of the compartmental contents will suffice. In others, of course, a determination of the entire contents of the compartment is required since the relevant quantities are "extensive" in nature. Often, however, one may consider the compartment as having been effectively entered, provided the necessary data concerning it can be known, if not by direct determination, by inference. One typical example is in a two-compartment system when only one compartment is initially labeled. Here it may be inferred that there is no activity in compartment 2 at the beginning of an experiment.

Often, the assumption of zero initial activity in all compartments can be asserted with confidence. If, in the two-compartment case, the experimenter now introduces a known amount of activity into an initially clean system, he can always infer the contents of the second compartment at any time from the difference between the remaining amount in compartment 1 and the initial amount. Assurance, of course, is required that there are no other compartments into which the label can penetrate.

Inferences concerning the total amount of S in a system are not always easily obtained. Nevertheless, if the specific activity of the initially labeled compartment ultimately stabilizes, it may perhaps be inferred that this terminal figure represents the overall specific activity in the system. In this case, S may be determined from simple isotope dilution (18). In a two-compartment system if one sample can be obtained from the second compartment at some time during the experiment, then further observations in compartment 1 alone will be sufficient to determine S_1 and S_2. If we know both S and R, we can then obtain the required a's.

Other inferences which often enlarge the scope of tracer experiments are: the assumption that S moves between compartments by pure exchange, the assumption that transport processes are constant, and finally, the important assumption that there are restrictions on communications between certain compartments. Although these assumptions may perhaps appear obvious in general, we must devote space to them because, in some instances, they have been taken for granted without adequate proof.

Constrained Systems

Few of the systems which are of interest will involve communication among all compartments without restriction. The imposition of constraints

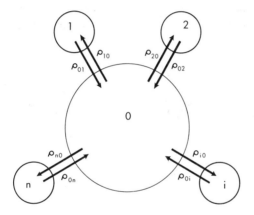

Fig. 11. A mammillary system showing the central compartment (numbered zero) and peripheral compartments numbered from 1 to n and including compartment i.

upon the general multicompartment system will, by reducing the number of ρ's which must be determined, also reduce the number of equations required. One type of constraint might be that all rates into and out of one or more particular compartments are zero. In this case it will usually be possible to obtain a solution for a system of fewer compartments requiring the use of fewer tracers or experiments.

Compartments which actually represent physical regions in space will, of course, be subject to the restriction that no compartment can communicate with any other which is not its neighbor. Of the systems subject to this rule two important ones can be considered. The first is the mammillary system (Fig. 11). Here one central compartment is surrounded by a series of peripheral ones. The other is the catenary system, in which compartments are arranged in series (Fig. 12). Of course, a closed two-compartment system can be regarded equally from either viewpoint. Since the most

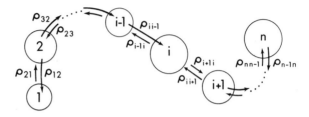

Fig. 12. A catenary system showing serially arranged compartments from 1 to n including the ith compartment.

interesting situation in the mammillary system is the one where the central compartment plays the predominant role and is initially labeled, a three compartment system may be considered as catenary if an end compartment

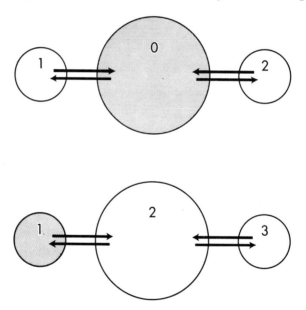

Fig. 13. Alternative initial labeling in a three-compartment system.

is initially labeled and as mammillary if the central compartment is (Fig. 13). Actually, however, there is nothing about the present analysis which requires the end compartment of a catenary system to be the initially labeled one.

It will be noted at once that in both of these systems there are only $2(n - 1)$ transport rates to be determined. Thus, we may reduce the required number of equations to this amount, and, at the same time, the required number of tracers or experiments. Normally such systems can be studied by the use of one labeled species of S and determinations of total amounts of S in the various compartments. In the closed mammillary system the central compartment can most conveniently be considered as compartment 0. In systems of this sort the analysis is simplified by the fact that observations are required only between neighboring compartments. Thus for the ith peripheral compartment,

$$\rho_{i0} a_0 - \rho_{0i} a_i = \frac{dR_i}{dt}$$

$$\rho_{i0} - \rho_{0i} = \frac{dS_i}{dt} \tag{25}$$

Since $dR_i/dt = d(a_i S_i)/dt = a_i\, dS_i/dt + S_i\, da_i/dt$, the solution of these equations yields

$$\rho_{i0} = \left[S_i \frac{da_i}{dt} \right] (a_0 - a_i)^{-1}$$

$$\rho_{0i} = \rho_{i0} - \frac{dS_i}{dt} \tag{26}$$

Observations of specific activities in the central and peripheral compartments and of S and rates of change of a and S in the peripheral compartment will suffice to determine the ρ's.

For the closed catenary system we consider compartment i, which communicates with its neighbor $i + 1$. We also consider a boundary between the compartments. Here, in addition to determinations of specific activity in the adjoining compartments i and $i + 1$, it is necessary to know the total flow of activity and of S between them. Usually this information will require a knowledge of the rate of change of R and S in all compartments on one side or the other of the boundary. These rates may be represented respectively as $\sum_{i+1}^{n} dR_j/dt$ and $\sum_{i+1}^{n} d(S_j/dt)$, where summation is over the running index j from compartment $i + 1$ to the last one, n. The total activity crossing the boundary from i to $i + 1$ (Fig. 12) is the difference between that moving from left to right and that moving oppositely.

Thus

$$a_i \rho_{i+1,i} - a_{i+1} \rho_{i,i+1} = \sum_{i+1}^{n} \frac{dR_j}{dt} \tag{27}$$

Similarly, for total S crossing the boundary

$$p_{i+1,i} - p_{i,i+1} = \sum_{i+1}^{n} \frac{dS_j}{dt} \tag{28}$$

The two equations may be solved for the p's yielding

$$p_{i+1,i} = \frac{a_{i+1} \sum\limits_{i+1}^{n} dS_j/dt - \sum\limits_{i+1}^{n} dR_j/dt}{a_{i+1} - a_i}$$

$$p_{i,i+1} = \frac{a_i \sum\limits_{i+1}^{n} dS_j/dt - \sum\limits_{i+1}^{n} dR_j/dt}{a_{i+1} - a_i} \tag{29}$$

The terms such as $\sum\limits_{i+1}^{n} dR_j/dt$ represent a flux of activity which might conceivably be determined across the boundary under certain favorable conditions without entry into the compartments beyond $i + 1$.

4

Kinetic Equations for Compartmental Systems: The Mammillary System

Model Systems

Many instances can be found in the literature where idealized compartmental model systems have been constructed in attempts to interpret the kinetic results of tracer experiments. References to a few selected examples are given in the Selected Bibliography of Experimental Studies. It is not the purpose of this volume to assess the validity of particular models since they involve a rather wide diversity of fields. Judgements cannot be trusted unless accompanied by expert knowledge in the particular field of application of the model. The reader who is concerned with the "practical" questions related to tracer kinetics will be aided by several review papers (19, 20, 21, 22, 23, 24).

Discussions of tracer kinetics will also be found in books on tracer methodology or allied subjects (25, 26, 27, 28). The model system will frequently be a valuable adjunct to the armamentarium of the cautious experimenter, and at times may represent an adequate picture of the behavior of the system. One purpose of the present volume is to provide a clearer picture of the ways in which such models can yield good or bad information, and to prepare the investigator against the wrong conclusions which can so easily result from misuse of models.

A Selected Bibliography of Experimental Studies

The following studies contain applications of theoretical equations to data. This list of references from the literature should not be considered complete, but the articles listed will illustrate the present interrelation of theory and experiment in the tracer field.

(The numbers in this list have nothing to do with the reference numbers that appear in the text.)

1. Baker, N., R. A. Shipley, R. E. Clark, and G. E. Incefy, "C[14] Studies in Carbohydrate Metabolism: Glucose Pool Size and Rate of Turnover in the Normal Rat," *Am. J. Physiol.* **196**: 245–252, 1959.
2. Berson, S. A., and R. S. Yalow, "Quantitative Aspects of Iodine Metabolism: The Exchangeable Organic Iodine Pool and the Rates of Thyroidal Secretion, Peripheral Degradation and Fecal Excretion of Endogenously Synthesized Organically Bound Iodine," *J. Clin. Invest.* **33**: 1533–1552, 1954.
3. Berson, S. A., and R. S. Yalow, "The Distribution of I[131] Labeled Human Serum Albumin Introduced into Ascitic Fluid: Analysis of the Kinetics of a Three Compartment Catenary Transfer System in Man and Speculations on Possible Sites of Degradation," *J. Clin. Invest.* **33**: 377–387, 1954.
4. Burch, George E., Sam A. Threefoot, and James A. Cronvich, "Theoretic Considerations of Biologic Decay Rates of Isotopes," *J. Lab. Clin. Med.* **34**: 14–30, 1949.
5. Ginsburg, J. M., and W. S. Wilde, "Distribution Kinetics of Intravenous Potassium," *Am. J. Physiol.* **179**: 63–75, 1954.
6. Jardetzky, C. D., and C. P. Barnum, "Estimation of Metabolic Reaction Rates from a Mathematical Analysis of Tracer Experiments," *Arch. Biochem. Biophys.* **73**: 435–450. 1958.
7. Keynes, R. D., and P. R. Lewis, "The Resting Exchange of Radioactive Potassium in Crab Nerve," *J. Physiol.* **113**: 73–98, 1951.
8. Matthews, Christine M. E., "The Theory of Tracer Experiments with I[131]-Labelled Plasma Proteins," *Phys. Med. Biol.* **2**: 36–53, 1957.
9. McLennan, H., "The Transfer of Potassium Between Mammalian Muscle and the Surrounding Medium," *Biochim. et Biophys. Acta* **16**: 87–95, 1955.
10. Oddie, T. H., "Analysis of Radio-Iodine Uptake and Excretion Curves," *Brit. J. Radiol.* **22**: 261–267, 1949.
11. Olsen, Norman S., and G. G. Rudolph, "Transfer of Sodium and Bromide Ions Between Blood, Cerebrospinal Fluid and Brain Tissue," *Am. J. Physiol.* **183**: 427–432, 1955.
12. Riggs, D. S., "Quantitative Aspects of Iodine Metabolism in Man," *Pharmacol. Revs.* **4**: 284–370, 1952.
13. Wang, J. C., "Penetration of Radioactive Sodium and Chloride into Cerebrospinal Fluid and Aqueous Humor," *J. Gen. Physiol.* **31**: 259–268, 1948.

A number of papers, supplementing those on practical tracer theory, have appeared in recent years; these treat the subject from a general non-specific viewpoint and emphasize the basic mathematical principles with little concern for some particular laboratory situation (4, 29, 30, 31, 32, 33, 34, 35). Contention as to whether the "basic" or "practical" point of view has the greatest merit would seem to be rather pointless. The reader will consult papers from one or both categories, depending on his interest and needs. One publication which is particularly valuable in bridging this gap is the well-thought-out discussion by Wrenshall (36). Although not devoted to the subject of radioactive tracers, two early papers by Teorell (37) are of historical interest. Much of their subject matter is of direct application to tracer theory particularly in physiological systems.

The model approach generally postulates some system of uniformly mixed compartments which is believed to resemble the system under study. Transport rates are postulated between the compartments, differential equations are set up and curves of specific activity in the compartments are mathematically predicted by integrating the equations. The results are then tested against data obtained in the actual system and a decision is made concerning the satisfactory or unsatisfactory nature of the comparison. Part of the strategy in the use of such a method lies in the art of selection among various competing models, particularly in the presence of fluctuation in the data. It must be admitted that all too frequently in past work the amounts of fluctuation have been too great in some physiological systems to permit unique choices to be made.

Mathematically speaking, it is fortunate that the behavior of tracers in steady-state multicompartment systems can be formulated in terms of systems of linear first-order differential equations with constant coefficients. Such equations are among the not-too-common types for which general solutions may be assured by straightforward methods. One example has already been cited in Chapter 1 for the two-compartment system. Before we proceed to more general systems, it will be necessary to devote a few pages to an outline of mathematical methods. An elementary knowledge of integral calculus and considerable facility with algebra must be assumed (38). Nevertheless the reader who can be satisfied with a knowledge of the mechanics of solution from a special viewpoint without mathematical rigor, should experience little difficulty once familiarity has been gained with a few basic concepts.

Use of the Laplace Transform

The method which we are about to consider is analogous in certain respects to the use of logarithms for obtaining the product of two numbers. In order to multiply, we employ a table in which the logarithm of each factor is found; the logarithms are added; and the resulting antilogarithm is then found by reference again to the table. At this point we may consider that the number has been "transformed" into its logarithm. After a simple operation is completed on the transformed quantities the final result is obtained by an "inverse transformation."

In the case of a linear differential equation the obstacle to its solution is the presence of a derivative in the equation. In the simpler differential equations, derivatives can be eliminated by a Laplace transformation. The resulting algebraic equation is solved by simple methods, and the final solution is subjected to an inverse transformation or "inversion," yielding the desired solution of the differential equation. Here, of course, we are transforming functions rather than simple numbers. These methods have had widespread use in engineering and applied mathematics (39). They are useful for the most part in the case of linear equations with constant coefficients. In other cases, obtaining the transforms or inverting them will usually prove difficult.

The Laplace transform of a function of t (assumed transformable) is represented by the notation $\mathscr{L}[f(t)]$. It can be obtained by multiplying the function $f(t)$ by e^{-st} and integrating over t from zero to infinity. Thus we define the Laplace transform as follows:

$$\mathscr{L}[f(t)] = \int_0^\infty e^{-st} f(t)\, dt = F(s) \tag{1}$$

Since t becomes a dummy variable, it disappears in the integration, and a new independent variable s appears in its place. This variable is a complex number but, in what follows, can be handled algebraically just as any other number. Lower-case notation is used in some books for the original function and upper case for the transformed function. In other instances, transforms are indicated by the use of the corresponding Greek letter. Inversion is represented as

$$\mathscr{L}^{-1}[F(s)] = f(t)$$

A few simple rules which establish relations between functions and their transforms are as follows:

A. If $\mathscr{L}[f_1(t)] = F_1(s)$ and $\mathscr{L}[f_2(t)] = F_2(s)$,

then
$$\mathscr{L}[f_1(t) + f_2(t)] = F_1(s) + F_2(s) \tag{2}$$

B. If K is a constant,

$$\mathscr{L}[Kf(t)] = K\mathscr{L}[f(t)] = KF(s) \tag{3}$$

C.
$$\mathscr{L}\left[f\left(\frac{t}{a}\right)\right] = aF(as) \tag{4}$$

D.
$$\mathscr{L}\left[\frac{df}{dt}\right] = sF(s) - f(0)^+ \tag{5}$$

where $f(0)^+$ is the value of $f(t)$ at $t = 0$.

E.
$$\lim_{t \to \infty} f(t) = \lim_{s \to 0} sF(s) \tag{6}$$

F.
$$\mathscr{L}\left[\int_0^t f(t)\, dt\right] = \frac{F(s)}{s} \tag{7}$$

Relation A is an expression of the additivity of functions and their transforms. The transform of a sum is the sum of the transforms.

Relation B expresses the fact that, if a constant is factored out of $f(t)$, it may be factored out of the transform. According to relation C, if a scale change is made in the independent variable t, an inverse change will occur in the transform both in the independent variable and in the overall scale of $F(s)$. Relation D permits the transform of the derivative of $f(t)$ to be expressed in terms of the transform of $f(t)$. By $f(0)^+$ is meant the value of f as the function approaches $t = 0$ from the *positive* side, since under certain circumstances the analysis can be applied to functions which jump discontinuously at $t = 0$. Relation E expresses the fact that, as t becomes very large in $f(t)$, and s becomes very small in $F(s)$, the function and s times its transform approach one another (provided the limiting values of both exist). Relation F expresses a relation between a transform of a function and the integral of the function. A further important theorem relating to the products of transforms and typical application will be found in Chapter 5 (pp. 85–87). Although many more important relations similar to those above can be written, they will not be required in the present analysis.

In using the Laplace transform method one may be greatly aided by tables of functions and their transforms. A number are available in the literature (40), but in certain cases transforms will be found which have been defined to be s times as great as those employed here. Table 1 provides a brief list of transforms which may be encountered in tracer kinetics. The Laplace variable s will always be written in lower case and must be distinguished from the upper-case S which continues to represent "substance."

Application to the two-compartment system. Consider in a simple closed two-compartment steady-state system the equation

$$d\Delta/dt = -\rho(1/S_1 + 1/S_2)\Delta \tag{8}$$

where Δ is the difference in specific activities between the compartments. The solution by Laplace transforms is unnecessary in this case (separable equation) and was already obtained in Chapter 1, but it has considerable illustrative value. We denote the transform of Δ as $D(s)$. At $t = 0$ if one compartment alone is initially labeled with specific activity $a(0)$ the difference Δ in specific activities of the two compartments is $\Delta(0) = a(0)$. If the transform of both sides of 8 is taken,

$$\mathscr{L}\frac{d\Delta}{dt} = s\,D(s) - a(0) = \mathscr{L}\left[-\rho\left(\frac{1}{S_1} + \frac{1}{S_2}\right)\Delta\right]$$

$$= -\rho\left(\frac{1}{S_1} + \frac{1}{S_2}\right)D(s) \tag{9}$$

Table I

*Some Laplace-transformable functions and their transforms
which are useful in tracer kinetics*

Constants, such as a_0 and δ, are used in this table as general arbitrary quantities which relate functions to their transforms. No relation should be assumed to the notation employed elsewhere in the book.

Function	Transform
$f(t)*$	$F(s)$
$e^{-\gamma t}$	$\dfrac{1}{s + \gamma}$
$\dfrac{1 - e^{-\gamma t}}{\gamma}$	$\dfrac{1}{s(s + \gamma)}$
$\dfrac{e^{-\gamma t} - e^{-\delta t}}{\delta - \gamma}$	$\dfrac{1}{(s + \gamma)(s + \delta)}$
$\dfrac{(a_0 - \delta)e^{-\delta t} - (a_0 - \gamma)e^{-\gamma t}}{\gamma - \delta}$	$\dfrac{s + a_0}{(s + \gamma)(s + \delta)}$
$\dfrac{1}{\gamma\delta} + \dfrac{\gamma e^{-\delta t} - \delta e^{-\gamma t}}{\gamma\delta(\delta - \gamma)}$	$\dfrac{1}{s(s + \gamma)(s + \delta)}$
$\dfrac{a_0}{\gamma\delta} + \dfrac{(a_0 - \gamma)e^{-\gamma t}}{\gamma(\gamma - \delta)} + \dfrac{(a_0 - \delta)e^{-\delta t}}{\delta(\delta - \gamma)}$	$\dfrac{s + a_0}{s(s + \gamma)(s + \delta)}$
$\dfrac{a_0}{\gamma\delta} + \dfrac{(\gamma^2 - a_1\gamma + a_0)e^{-\gamma t}}{\gamma(\gamma - \delta)} + \dfrac{(\delta^2 - a_1\delta + a_0)e^{-\delta t}}{\delta(\delta - \gamma)}$	$\dfrac{s^2 + a_1 s + a_0}{s(s + \gamma)(s + \delta)}$
$\dfrac{t^n e^{-\alpha t}}{n!}$	$\dfrac{1}{(s + \alpha)^{n+1}}$
$\dfrac{e^{-\alpha t}}{\beta} \sin \beta t$	$\dfrac{1}{(s + \alpha)^2 + \beta^2}$
$\dfrac{e^{-\alpha t}}{\beta} \sinh \beta t$	$\dfrac{1}{(s + \alpha)^2 - \beta^2}$

* Whenever $f(t)$ contains $e^{-\lambda t}$, the exponential expression is taken to be zero for all negative t.

Solving for D, we have

$$D(s) [s + \rho(1/S_1 + 1/S_2)] = a(0)$$
$$D(s) = a(0)/[s + \rho(1/S_1 + 1/S_2)] \tag{10}$$

To obtain the function $\Delta(t)$ corresponding to transform $D(s)$, we consult line 2 of Table 1. Any function of the form $F(s) = 1/(s + \gamma)$ is the transform of $f(t) = e^{-\gamma t}$. Here, corresponding to γ, we have $\rho(1/S_1 + 1/S_2)$.

According to relation B (p. 50), a multiplier $a(0)$ of the transform is an identical multiplier of the function. Thus the inverse of D is given by

$$f(t) = \mathscr{L}^{-1}[D(s)] = \Delta = a(0)e^{-\rho(1/S_1 + 1/S_2)t} \tag{11}$$

In general, to invert a transform by the use of tables, we proceed algebraically to attempt to throw it into some form whose inverse is to be found in the tabular list of transforms.

General Kinetic Equations for a Closed Multicompartment System

In determining the general kinetic response of a multicompartment system it is preferable to choose one dependent variable for each compartment. The independent variable will be the time t. Relations 2 and 3, in Chapter 3, have been previously established by equating the income and outgo in a general compartment. Thus

$$\frac{dR_i}{dt} = \frac{d(a_i S_i)}{dt} = \sum_j (\rho_{ij} a_j - a_i \rho_{ji})$$

$$dS_i \, dt = \sum_j (\rho_{ij} - \rho_{ji})$$

Insert for $d(a_i S_i)/dt$ the equivalent $a_i dS_i/dt + S_i \, da_i/dt$ on the left side of the first expression, and multiply the second by a_i. Subtract the first from the second, and divide by S_i; the result will be

$$\frac{da_i}{dt} = \sum_j \frac{\rho_{ij}}{S_i}(a_j - a_i) \tag{12}$$

This equation will be the general kinetic relation for a multicompartment system, whether or not the system is in a steady state. The ρ's and S's can also be functions of t. We may conclude in general that the kinetic response of any multicompartment system does not depend specifically on the ρ's, but on the ratios of the ρ's to the compartmental contents.

Equation 12 forms a set of differential-difference equations. They are differential equations because the derivative of the a's is related to the a's themselves. They are difference equations because the a's in various compartments are related to their intercompartmental differences. Solutions of the system for a given set of initial conditions relate the a's to the values of ρ/S and to the time, and they also establish the kinetic behavior of the physical system they describe. Since the behavior of a system is determined by the values of ρ/S for the various compartments, it is clear that all ρ's and S's will enter into the determination. The initial values $a_i(0)$ will also be involved. These values are the initial specific activities of all compartments.

In general the rate of change of a_i, depending as it does on the difference between the local a and the a's for all other compartments, represents a kind of overall "driving force" in each compartment. If all a's are alike, the derivatives vanish, and there will be no further change in the a's, since the right hand side of eq. 12 is zero. Therefore, if an equilibrium state is reached in the system, it will persist from then on, and the a's will thereafter be constant. We might be tempted to consider a mathematical proof that the a's cannot oscillate under any possible physical conditions which do not involve the introduction of further activity into the system or the separation of isotopes. The conclusion will preferably rest on the physical principle that the entropy of a system cannot spontaneously decrease. From the physical point of view it is also clear that the final equilibrium value of the a's will be given by

$$a_i(\infty) = \frac{\sum_{i=1}^{n} a_i(0)S_i(0)}{S} \tag{13}$$

where S is the total amount of substance in the system.

Constant coefficients.   At present there is no practical procedure by which these equations can be analytically solved in the general case. When the quotients ρ/S are all constant, the method of Laplace transforms can be applied. In a general system of n compartments without constraints, we can take the transform of both sides of eq. 12. For $\mathscr{L}(a_i)$, we will use the symbol α_i. For the right-hand side, we can express the transform of the sum as the sum of the transforms of the individual members term by term from relation 2. Also, because of relation 3, the constants ρ/S can be factored out. For the transform of the left side we use relation 5. As a result we obtain

$$\mathscr{L}\left[\frac{d(a_i)}{dt}\right] = s\alpha_i - a_i(0) = \sum_{j} \frac{\rho_{ij}}{S_i}(\alpha_j - \alpha_i)$$

or

$$-\sum_{j}\left(\frac{\rho_{ij}}{S_i}\alpha_j\right) + \left(s + \sum_{j}\frac{\rho_{ij}}{S_i}\right)\alpha_i = a_i(0) \tag{14}$$

In this expression the first summation involves the α_j, but the second does not involve α_i. As before, solutions can be obtained in determinant form, and, by the appropriate use of tables of transforms, the α's can be inverted. Under rather broad conditions, inversion of expressions of this sort, involving denominators which are products of a finite number of factors, can be obtained by straightforward methods (39). The results of the inversion will be linear sums of a series of exponential factors multiplied by constants. A few selected examples will be discussed in this and the following chapter.

Kinetics of Mammillary Systems

The mammillary system has been discussed in some detail previously (4), and its study is advantageous for illustrating principles and methods of general compartmental kinetic analysis. As before, we will use the Laplace transform method. The transforms for a mammillary system can be related by inserting into eq. 14 the correct values of ρ_{ij}, but it is more instructive to begin with the appropriate differential equations. We will reserve the index zero for the central compartment. The number of *peripheral* compartments will be n. Because of the methodological restriction to a steady-state system, we postulate constant exchange rates between compartments. Thus $\rho_{0i} = \rho_{i0}$, which we will call ρ_i. For the peripheral and central compartments, then, we will have

$$\frac{da_0}{dt} = \sum_i \frac{\rho_i}{S_0} (a_i - a_0)$$

$$\frac{da_i}{dt} = \frac{\rho_i}{S_i} (a_0 - a_i) \tag{15}$$

Taking Laplace transforms, we obtain

$$s\alpha_0 - a_0(0) = \sum_i \frac{\rho_i}{S_0} (\alpha_i - \alpha_0)$$

$$s\alpha_i - a_i(0) = \frac{\rho_i}{S_i} (\alpha_0 - \alpha_i) \tag{16}$$

Writing the equations out will indicate the order which should be preserved for convenience in obtaining solutions. Thus in a system of n peripheral compartments we have the following equations:

$$\left(s + \sum_i \frac{\rho_i}{S_0}\right)\alpha_0 \quad - \frac{\rho_1}{S_0}\alpha_1 - \frac{\rho_2}{S_0}\alpha_2 - \cdots \frac{\rho_n}{S_0}\alpha_n = a_0(0)$$

$$-\frac{\rho_1}{S_1}\alpha_0 + \left(s + \frac{\rho_1}{S_1}\right)\alpha_1 \quad + 0 \quad \cdots \quad\quad = a_1(0)$$

$$\cdots\cdots\cdots\cdots\cdots\cdots\cdots\cdots\cdots\cdots\cdots\cdots\cdots\cdots\cdots$$

$$-\frac{\rho_n}{S_n}\alpha_0 \quad\quad + 0 \cdots + \left(s + \frac{\rho_n}{S_n}\right)\alpha_n = a_n(0)$$

From this set we may proceed to evaluate the determinants. A general solution will be unnecessarily cumbersome, so it will usually be of more interest to obtain particular solutions for cases of interest. In one such case all compartments but the central one are initially unlabeled. Therefore, inserting $a_i(0) = 0$, for $i \neq 0$, into the determinants, we obtain the following expression for α_0:

$$\alpha_0 = \frac{\begin{vmatrix} a_0(0) & -\rho_1/S_0 & \cdots & -\rho_n/S_0 \\ 0 & (s + \rho_1/S_1) & \cdots & 0 \\ 0 & 0 & (s + \rho_2/S_2) & \cdots & 0 \\ \cdot & \cdot & \cdot & \cdot \\ \cdot & \cdot & \cdot & \cdot \\ \cdot & \cdot & \cdot & \cdot \\ 0 & \cdots 0 \cdots & 0 & \cdots (s + \rho_n/S_n) \end{vmatrix}}{\begin{vmatrix} \left[s + \sum_i (\rho_i/S_0)\right] & -\rho_1/S_0 & -\rho_2/S_0 & \cdots & -\rho_n/S_0 \\ -\rho_1/S_1 & (s + \rho_1/S_1) & 0 & \cdots & 0 \\ -\rho_2/S_2 & 0 & (s + \rho_2/S_2) & \cdots & 0 \\ \cdot & \cdot & \cdot & \cdot \\ \cdot & \cdot & \cdot & \cdot \\ \cdot & \cdot & \cdot & \cdot \\ -\rho_n/S_n & 0 & 0 & \cdots & (s + \rho_n/S_n) \end{vmatrix}} \quad (17)$$

From theorems 3 and 4 (p. 35), we see that the numerator is

$$a_0(0) \times \left(s + \frac{\rho_1}{S_1}\right) \times \cdots \left(s + \frac{\rho_n}{S_n}\right) = a_0(0) \prod_{i=1}^{i=n}\left(s + \frac{\rho_i}{S_i}\right) \quad (18)$$

The use of the generalized product symbol Π is analogous to the use of Σ for generalized sums.

To evaluate the denominator, we first replace the elements of column 1 by the sum of elements of all columns (theorem 2, p. 35). This column will now be a series of s's, which, after we factor out s (theorem 6, p. 36), leaves 1's. Next $(s + \rho_i/S_i)$ is factored out of column $i + 1$ for each column in turn. Columns 2, 3, etc., are all subtracted from column 1.

Thus, under the conditions of theorems 3 and 4 (p. 35), the determinant becomes:

$$\Delta = s\left(s + \frac{\rho_1}{S_1}\right) \cdots \left(s + \frac{\rho_n}{S_n}\right) \begin{vmatrix} 1 + \sum_i \dfrac{\rho_i}{S_0(s + \rho_i/S_i)} & -\dfrac{\rho_1}{S_0(s + \rho_1/S_1)} & \cdots & -\dfrac{\rho_n}{S_0(s + \rho_n/S_n)} \\ 0 & 1 & \cdots & 0 \\ 0 & 0 & \cdots & 0 \\ 0 & 0 & \cdots & 0 \\ \cdots\cdots\cdots\cdots\cdots\cdots & & & \\ 0 & \cdots & \cdots & 1 \end{vmatrix}$$

$$= s\left[\prod_{i=1}^{i=n} \frac{s + \rho_i}{S_i}\right]\left[1 + \sum_{i=1}^{i=n} \frac{\rho_i}{S_0(s + \rho_i/S_i)}\right] \tag{19}$$

Division of the numerator (eq. 18) by Δ yields

$$\alpha_0 = \frac{a_0(0)}{s\left\{1 + \sum\limits_{i=1}^{i=n} \rho_i/[S_0(s + \rho_i/S_i)]\right\}} = \frac{a_0(0)}{sL(s)} \tag{20}$$

By rule F (eq. 7), we may remove a factor s from the denominator if the final inverted result is integrated between 0 and t. Also, by rule B (eq. 3), we may factor out the constant $a_0(0)$ and multiply the result by it. Thus we can invert the transform if we can invert the reciprocal of

$$L(s) = \left\{1 + \sum_i \rho_i/[S_0(s + \rho_i/S_i)]\right\} \tag{21}$$

Solution for a Three-Compartment Mammillary System with Central Compartment Initially Labeled

The central compartment. The inversion of α_0 for a three-compartment mammillary system will provide an instructive illustration of the general principle. In this case, writing $L(s)$ for $n = 2$, we have

$$\frac{\alpha_0}{a_0(0)} = \left\{s\left[1 + \frac{\rho_1}{S_0(s + \rho_1/S_1)} + \frac{\rho_2}{S_0(s + \rho_2/S_2)}\right]\right\}^{-1} \tag{22}$$

In relatively simple instances such as this one, if the transform can be expressed as a rational proper fraction, its equivalent can usually be recognized in the transform tables. To do so in the present situation we may simplify the notation by choosing for the fractional exchange rates $\rho_i/S_i = \beta_i$ and $\rho_i/S_0 = B_i$. Thus, if we reduce $L(s)$ to a common denominator and take the reciprocal,

$$\frac{\alpha_0}{a_0(0)} = \frac{(s + \rho_1/S_1)(s + \rho_2/S_2)}{s[(s + \rho_1/S_1)(s + \rho_2/S_2) + (\rho_1/S_0)(s + \rho_2/S_2) + (\rho_2/S_0)(s + \rho_1/S_1)]}$$

$$= \frac{(s + \beta_1)(s + \beta_2)}{s[(s + \beta_1)(s + \beta_2) + B_1(s + \beta_2) + B_2(s + \beta_1)]}$$

The quadratic expression in the denominator may be factored into $(s + \lambda_1)(s + \lambda_2)$ where

$$-\lambda_1 = \frac{-(\beta_1 + \beta_2 + B_1 + B_2) - D^{1/2}}{2}$$

$$-\lambda_2 = \frac{-(\beta_1 + \beta_2 + B_1 + B_2) + D^{1/2}}{2} \tag{23}$$

and $\quad D = (\beta_1 - \beta_2)^2 + (B_1 + B_2)^2 + 2(\beta_1 - \beta_2)(B_1 - B_2). \tag{24}$

Since $-\lambda_1$ and $-\lambda_2$ are the roots of the denominator, we obtain

$$\alpha_0 = a_0(0) \frac{(s + \beta_1)(s + \beta_2)}{s(s + \lambda_1)(s + \lambda_2)} \tag{25}$$

which, from Table 1, we recognize as the transform of

$$a_0(t) = a_0(0) \left[\frac{\beta_1\beta_2}{\lambda_1\lambda_2} + \frac{(\lambda_1 - \beta_1)(\lambda_1 - \beta_2)e^{-\lambda_1 t}}{\lambda_1(\lambda_1 - \lambda_2)} + \frac{(\lambda_2 - \beta_1)(\lambda_2 - \beta_2)e^{-\lambda_2 t}}{\lambda_2(\lambda_2 - \lambda_1)} \right] \tag{26}$$

It may be noted that $-\lambda_{1,2}$ are the roots obtained from the relation $L(s) = 0$.

We recall from elementary algebra that the product of the roots of a quadratic expression is the term without s, which in the present case is $\beta_1\beta_2 + B_1\beta_2 + B_2\beta_1$. The first term in the bracket of eq. 26 represents the specific activity at an infinite time when the tracer is completely mixed in the system. This expression may be evaluated as

$$a(\infty) = a_0(0) \frac{\beta_1\beta_2}{\beta_1\beta_2 + B_1\beta_2 + B_2\beta_1}$$

$$= \frac{a_0(0)}{(1 + B_1/\beta_1 + B_2/\beta_2)}$$

$$= \frac{a_0(0)}{(1 + S_1/S_0 + S_2/S_0)}$$

$$= a_0(0) \frac{S_0}{S} \tag{27}$$

This equation may be compared with the more general eq. 13, which was predicted on physical grounds alone.

The peripheral compartment.   The utility of the transform method is particularly well shown in obtaining the solution for peripheral compartments. Here we have from eq. 16 in general,

$$\alpha_i = [a_i(0) + \beta_i \alpha_0](s + \beta_i)^{-1}$$

In the present case, where the initial specific activity is zero,

$$\alpha_i = \frac{\beta_i\,\alpha_0}{s + \beta_i} \tag{28}$$

In solving for α_i in the two peripheral compartments, we use eq. 25, which yields

$$\alpha_1 = a_0(0)\,\frac{\beta_1(s + \beta_2)}{s(s + \lambda_1)(s + \lambda_2)}$$

whose inverse is found in the table to be

$$a_1(t) = a_0(0)\left[\frac{\beta_1\beta_2}{\lambda_1\lambda_2} + \frac{(\beta_1\beta_2 - \beta_1\lambda_1)e^{-\lambda_1 t}}{\lambda_1(\lambda_1 - \lambda_2)} + \frac{(\beta_1\beta_2 - \beta_1\lambda_2)e^{-\lambda_2 t}}{\lambda_2(\lambda_2 - \lambda_1)}\right] \tag{29}$$

The solution for the second compartment is obtained by interchanging 1 and 2.

Lumping of peripheral compartments.   The kinetics of the central compartment exhibit an interesting property if the two peripheral compartments approach one another in their behavior. Here we can express the difference between them in terms of $\beta_1 - \beta_2$. As the exchange rates approach one another, we may consider that $\beta_1 - \beta_2$ and $B_1 - B_2$ become small. Since these differences appear as squared terms in eq. 24, then, as $\beta_1 \to \beta_2$ and $B_1 \to B_2$, $D \to (B_1 + B_2)^2$ very rapidly. As a result a good approximation of the λ's can be made as follows:

$$\lambda_1 = \frac{\beta_1 + \beta_2}{2} + B_1 + B_2$$

$$\lambda_2 = \frac{\beta_1 + \beta_2}{2}. \tag{30}$$

If we insert these values in eq. 26, the equation of the specific activity of the central compartment, we obtain for the coefficient of the second exponential term

$$\frac{(\beta_2 - \beta_1)^2}{2(\beta_1 + \beta_2)(B_1 + B_2)} \tag{31}$$

This term, which also contains the difference of the β's as a square, rapidly becomes negligible in its influence on the system kinetics as the β's approach one another. Consequently the dominating rate constant λ_1 controls the situation as though β_1 and β_2 acted only through their arithmetic mean. Two peripheral compartments must differ considerably in their behavior if, through observation in the central compartment alone, the system is to be recognized as containing three compartments.

General conclusions in a mammillary system with central compartment initially labeled: inversion by partial fractions. In the three-compartment case we recognize that, if there are two peripheral compartments ($n = 2$), there will be two exponential terms both in the solutions for the central compartment and in the solutions for the peripheral compartments. Furthermore the exponents will involve combinations of the properties of all compartments. Finally, for a closed system there will be a constant term. Similar conclusions can be extended to the general system. At the same time the process for the general case will be clarified by reference to the previously derived case where $n = 2$.

The clue to inversion of the transform for the general case is found in the previous analysis where the solutions were obtained in the form of a series of exponential terms. Recalling that such functions have transforms of the type $1/(s + \gamma)$, we may attempt to expand the transform of eq. 20 in a series of partial fractions. Details of the method will be found in standard textbooks (39). Here we will quote a necessary theorem without proof:

A theorem on partial fractions. Let any function of s be given which can be expressed as a quotient; the numerator $A(s)$ is a polynomial of degree n or less and the denominator $B(s)$, a polynomial of degree $n + 1$. Let the ith root of the denominator be $-\lambda_i$, where $1 \leq i \leq n + 1$. Then, if $B'(s)$ is the first derivative of B with respect to s,

$$\alpha(s) = \frac{A(s)}{B(s)} = \sum_{i=1}^{n+1} \frac{A(\lambda_i)}{B'(\lambda_i)(s + \lambda_i)} \tag{32}$$

$A(\lambda_i)$ and $B'(\lambda_i)$ are obtained by inserting λ_i in place of s in the original expressions. This is done for all λ_i's and summation is then performed for all λ_i's. The λ_i's are not variables but are constants determined by the particular form of $B(s)$. We obtain by this theorem a series of terms of the form $C_i/(s + \lambda_i)$, provided we may obtain the roots $-\lambda_i$ of $B(s)$. The partial fraction expansion is thus obtained by writing a sum of terms. One term is written for each root, the root being inserted in place of s in $A(s)$ and also in the first derivative of $B(s)$. The quotient of A/B' in each case thus evaluated is divided by $(s + \lambda_i)$. The inverse of this sum obtained from Table 1 is:

$$a(t) = \sum_i \frac{A(\lambda_i)}{B'(\lambda_i)} e^{-\lambda_i t} \tag{33}$$

We now apply these considerations to the general transform 20,

$$\alpha_0(s) = a_0(0)/sL(s),$$

where

$$L(s) = \left(1 + \sum_i \frac{B_i}{s + \beta_i}\right)$$

It is convenient to reduce $L(s)$ to a common denominator. Its numerator is then a polynomial of degree n. On taking the reciprocal this becomes the denominator of the transform so it must be expressed in terms of n roots $-\lambda_i$.

For symmetry we also introduce the root $\lambda_0 = 0$. There are now $n + 1$ roots, thus

$$\alpha_0(s) = a_0(0) \left[\frac{(s + \beta_1)(s + \beta_2) \cdots (s + \beta_n)}{(s + \lambda_0)(s + \lambda_1) \cdots (s + \lambda_n)}\right]$$

The partial fraction expansion is obtained from eq. 32 as

$$\alpha_0(s) = a_0(0) \sum_{i=0}^{i=n} \frac{X_i}{s + \lambda_i}$$

where the coefficients X_0, X_1, etc., are obtained by evaluating the numerator and the derivative of the denominator of $\alpha_0(s)/a_0(0)$ for $s = -\lambda_1$, etc.*
For the denominator

$$\frac{d}{ds}[(s + \lambda_0)(s + \lambda_1) \cdots (s + \lambda_n)] = (s + \lambda_1) \cdots (s + \lambda_n)$$

$$+ (s + \lambda_0)(s + \lambda_2) \cdots (s + \lambda_n) + \cdots + (s + \lambda_0) \cdots (s + \lambda_{n-1})$$

Whatever $-\lambda_i$ replaces s in this expression, all terms but the ith will vanish since this is the only term which does not include a factor containing $+\lambda_i$. Thus

$$X_i = - \left[\frac{(\beta_1 - \lambda_i)(\beta_2 - \lambda_i) \cdots (\beta_n - \lambda_i)}{\lambda_i(\lambda_1 - \lambda_i) \cdots (\lambda_{i-1} - \lambda_i)(\lambda_{i+1} - \lambda_i) \cdots (\lambda_n - \lambda_i)}\right]$$

* It is immaterial, of course, whether i ranges from 0 to n or from 1 to $n + 1$.

The transform for any peripheral compartment j is obtained by a similar analysis, and upon inversion according to eq. 33,

$$a_0(t) = a_0(0) \sum_{i=0}^{i=n} X_i e^{-\lambda_i t}$$

$$a_j(t) = a_0(0) \sum_{i=0}^{i=n} \left(\frac{\beta_j X_i}{\beta_j - \lambda_i}\right) e^{-\lambda_i t} \tag{34}$$

The term for $i = 0$ is of particular interest since $\lambda_0 = 0$. The corresponding exponential in the limit becomes a constant representing the final uniform, terminal specific activity. This is reached in the system after the kinetical processes have come to an end, with complete mixing of the tracer. For the case $s = \lambda_0 = 0$, we may obtain X_0 by evaluating $L(s)$ for $s = 0$. This becomes

$$L(0) = 1 + \sum_i \frac{B_i}{\beta_i} = 1 + \sum_i \frac{S_i}{S_0} = \frac{S}{S_0}$$

yielding $X_0 = S_0/S$ which may be predicted on physical grounds alone.

We may simplify eq. 34 slightly by writing

$$\frac{a_0(t)}{a_0(0)} = \frac{S_0}{S} + \sum_{i=1}^{i=n} X_i e^{-\lambda_i t}$$

$$\frac{a_j(t)}{a_0(0)} = \frac{S_0}{S} + \sum_{i=1}^{i=n} \left(\frac{\beta_j X_i}{\beta_j - \lambda_i}\right) e^{-\lambda_i t} \tag{35}$$

where

$$X_i = \frac{\prod\limits_{k=1}^{n} (\lambda_i - \beta_k)}{\prod\limits_{\substack{k=0 \\ k \neq i}}^{n} (\lambda_i - \lambda_k)} \tag{36}$$

for $i \neq 0$, and $-\lambda_i$ are the n roots of the equation

$$L(s) = 1 + \sum_{i=1}^{i=n} B_i/(s + \beta_i) = 0$$

The solution for this problem given by Sheppard and Householder (4) contains an error. The right hand side of their equation 12 should be multiplied through by α_j (using their notation) and the i and j in the denominator of their X_i should be interchanged.

Of course in determining the λ's, we must reduce $L(s)$ to a common denominator. This is possible for the purpose of determining the λ's because no λ is equal to any β_i. Since $L(s)$ appears in the denominator of the transform, it will be the numerator which must be equated to zero to obtain the roots. Because there are n factors $s + \beta_1$, the expression for the numerator will be a polynomial of degree n yielding n roots, and therefore n values of $-\lambda_i$. These plus $\lambda_0 = 0$ complete the solution. Since $L(s)$ is unaltered by interchanging any pair of subscripts i, neither the λ's nor the X_i's depend in any way on the order of the peripheral compartments. There is always a possibility that two roots will be equal, but this situation is unlikely in tracer experiments. For the procedure to

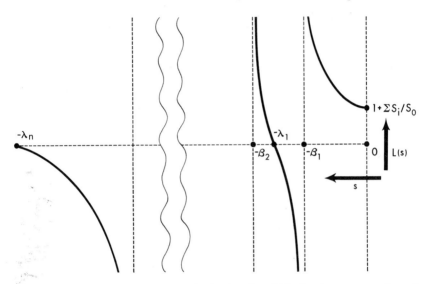

Fig. 14. Schematic representation of the function $L(s)$ showing roots $-\lambda_i$ and asymptotes $-\beta_i$.

be followed in this case, more advanced references must be consulted (39). In this general case, as in the case where $n = 2$, the λ's represent combinations of all of the β's and B's in the system. For $n \geq 5$ it will be necessary to obtain approximate values of the roots since the general equation of the fifth degree or higher has no exact solution.

The principle of lumping. In order to understand the effect of lumping of peripheral compartments in the generalized situation it is instructive to consider for a moment the graphical behavior of the function

$$L(s) = 1 + \sum_i \frac{B_i}{s + \beta_i}$$

as shown in Fig. 14. At $s = 0$ the function has the value $1 + \sum S_i/S_0$. Proceeding in the negative direction, the value will increase until s reaches $-\beta_1$ at which it will reach a positively infinite asymptote (see Fig. 14).

Passing beyond $-\beta_1$, it will reverse its sign and return from $-\infty$, increasing progressively until it reaches $-\beta_2$, where a second asymptote is situated. Between $-\beta_1$ and $-\beta_2$ the function must pass through zero at the first root $-\lambda_1$. In general it is evident that there is an asymptote for every β and that roots $-\lambda_i$ lie between $-\beta_i$ and $-\beta_{i+1}$.

Let it be supposed that $\beta_i \to \beta_{i+1}$ without any other change in the system. Then $-\lambda_i$ is constrained to lie between them so that $\lambda_i \to \beta_i \to \beta_{i+1}$. A situation is ultimately reached where the adjoining roots λ_{i-1} and λ_{i+1} are scarcely influenced by replacing β_i and β_{i+1} with their mean value $(\beta_i + \beta_{i+1})/2$. The remaining roots are influenced to an even less degree. If we examine the numerator X_i, we see that the two factors $(\lambda_i - \beta_i)$ and $(\lambda_i - \beta_{i+1})$ will both become small because λ_i is sandwiched in between the β's. The denominator will not be made small by this action since λ_i will still be well separated from its neighbors. Because there are two small factors in the numerator, the coefficient X_i will rapidly disappear and with it one of the exponential terms.

Therefore, here, as in the case $n = 2$, discussed on pages 59–60, the two similar peripheral compartments may be lumped into one so far as their effect on the rest of the system is concerned. By continuing this argument, we see that, unless compartments all differ materially in their behavior, groups of similar compartments will tend to act together, and the kinetics of the central compartment will be indistinguishable from the kinetics of a system of relatively few compartments. This principle is called the "principle of lumping." Arguments can also be constructed for the specific activity relations as observed in two or more similar peripheral compartments.

Further Analysis of a Mammillary System: One Peripheral Compartment Initially Labeled

The indices for all peripheral compartments are arbitrary, and it does not matter which is chosen as number 1. We may therefore choose it as the initially labeled compartment. In this case the transform for the central compartment is obtained from eq. 16 by inserting $a_i(0) = 0$ for $i \neq 1$ in the numerator determinant and evaluating in the usual manner. Thus

$$\alpha_0(s) = \frac{a_1(0)B_1}{s(s + \beta_1)L(s)} \tag{37}$$

where $L(s)$ is as previously defined in eq. 21.

Again we may express $L(s)$ as a quotient of polynomials, and

$$\alpha_0(s) = \frac{a_1(0)B_1(s + \beta_1)(s + \beta_2) \cdots (s + \beta_n)}{s(s + \beta_1)(s + \lambda_1)(s + \lambda_2) \cdots (s + \lambda_n)}$$

$$= a_1(0)B_1 \left[\frac{(s + \beta_2) \cdots (s + \beta_n)}{s(s + \lambda_1)(s + \lambda_2) \cdots (s + \lambda_n)} \right] \tag{38}$$

We next consider the transform of compartment 1, the initially labeled compartment. After evaluation of the numerator determinant using a method similar to that employed in the evaluation of the denominator of eq. 17, we may write the following expression:

$$\alpha_1(s) = a_1(0) \frac{s(s + \beta_2) \cdots (s + \beta_n) \left[1 + B_1/s + \sum\limits_{i=2}^{i=n} B_i/(s + \beta_i) \right]}{(s + \lambda_0) \cdots (s + \lambda_n)} \tag{39}$$

Similarly for other peripheral compartments than the initially labeled one,

$$\alpha_j(s) = a_1(0) \frac{B_1\beta_j(s + \beta_2) \cdots (s + \beta_{j-1})(s + \beta_{j+1}) \cdots (s + \beta_n)}{(s + \lambda_0) \cdots (s + \lambda_n)} \tag{40}$$

Partial fraction expansions are

$$\frac{\alpha_0(s)}{a_1(0)} = \frac{S_1}{Ss} + B_1 \sum_{i=1}^{n} \frac{X_i}{(s + \lambda_i)(\beta_1 - \lambda_i)}$$

$$\frac{\alpha_1(s)}{a_1(0)} = \frac{S_1}{Ss} + \sum_{i=1}^{n} \frac{\lambda_i X_i \left[1 - B_1/\lambda_i + \sum\limits_{j=2}^{j=n} B_j/(\beta_j - \lambda_i) \right]}{(\lambda_i - \beta_1)(s + \lambda_i)}$$

$$\frac{\alpha_j(s)}{a_1(0)} = \frac{S_1}{Ss} + B_1\beta_j \sum_{i=1}^{n} \frac{X_i}{(s + \lambda_i)(\beta_1 - \lambda_i)(\beta_j - \lambda_i)} \tag{41}$$

where $j \neq 1$.
Inverting we obtain

$$\frac{a_0(t)}{a_1(0)} = \frac{S_1}{S} + B_1 \sum_{i=1}^{n} \left(\frac{X_i}{\beta_1 - \lambda_i} \right) e^{-\lambda_i t}$$

$$\frac{a_1(t)}{a_1(0)} = \frac{S_1}{S} + \sum_{i=1}^{n} \left(\frac{\lambda_i X_i}{\lambda_i - \beta_1} \right) \left(1 - \frac{B_1}{\lambda_i} + \sum_{j=2}^{j=n} \frac{B_j}{\beta_j - \lambda_i} \right) e^{-\lambda_i t}$$

$$\frac{a_j(t)}{a_1(0)} = \frac{S_1}{S} + B_1\beta_j \sum_{i=1}^{n} \left[\frac{X_i}{(\beta_1 - \lambda_i)(\beta_j - \lambda_i)} \right] e^{-\lambda_i t} \tag{42}$$

where $j \neq 1$.
These expressions and those derived in the next chapter indicate the power and generality of the Laplace transform method. However, they

will involve rather cumbersome calculations if they are applied to actual systems with large numbers of compartments. They are not included here specifically for that purpose, but rather because the solutions for certain two-, three-, or four-compartment systems can be obtained from them by simple substitution into the expressions. By including one or more infinite compartments, we can also handle a few open systems in this way.

Practical Illustration of the Use of General Solutions

One practical illustration of the use of general solutions is the system shown in Fig. 15. Here we have a constant washout of an initially labeled central compartment which exchanges with a single neighbor compartment.

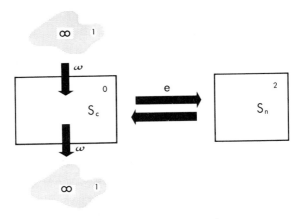

Fig. 15. Washout of a central compartment (0 in the figure) containing S_c units of substance. Exchange also occurs with a second compartment (2 in the figure) containing S_n units. Note how washout can be represented as exchange with an infinite compartment.

We recognize the "washout" as simply an exchange with an infinite compartment, which we will take as compartment 1 of a mammillary system. The central compartment will be compartment 0, and the neighbor compartment will be number 2.

The washout rate w, the exchange rate e, and the compartmental sizes S_c and S_n (see Fig. 15) can be combined so that we can substitute into general relations 35 and 36:

$$B_1 = w/S_c \qquad \beta_1 = w/\infty = 0$$
$$B_2 = e/S_c \qquad \beta_2 = e/S_n \qquad (43)$$

From these equivalents we must obtain the roots $-\lambda_{1,2}$ of

$$L(s) = 1 + \frac{w}{sS_c} + \frac{e}{S_c(s + e/S_n)} = 0$$

that is, $$s^2 + \left(\frac{e}{S_n} + \frac{w + e}{S_c}\right)s + \frac{ew}{S_cS_n} = (s + \lambda_1)(s + \lambda_2) = 0 \qquad (44)$$

where $\lambda_{1,2} = \left(\dfrac{e}{2S_n} + \dfrac{w + e}{2S_c}\right) \mp \dfrac{1}{2}\sqrt{\dfrac{e^2}{S_n^{\,2}} + \dfrac{2e(e - w)}{S_nS_c} + \dfrac{(w + e)^2}{S_c^{\,2}}}$ \qquad (45)

so that, if D is the quantity under the radical,

$$\lambda_1 = \left(\frac{e}{2S_n} + \frac{w + e}{2S_c}\right) + \frac{D^{1/2}}{2}$$

$$\lambda_2 = \left(\frac{e}{2S_n} + \frac{w + e}{2S_c}\right) - \frac{D^{1/2}}{2}$$

In computing the X's we employ

$$\lambda_1 - \lambda_2 = D^{1/2}, \quad \lambda_1 - \beta_1 = \lambda_1, \quad \lambda_2 - \beta_1 = \lambda_2$$

$$\lambda_1 - \beta_2 = \frac{w + e}{2S_c} - \frac{e}{2S_n} + \frac{D^{1/2}}{2}$$

$$\lambda_2 - \beta_2 = \frac{w + e}{2S_c} - \frac{e}{2S_n} - \frac{D^{1/2}}{2}$$

yielding $X_0 = 0$,

$$X_1 = \frac{1}{2} + \left(\frac{w + e}{2S_c} - \frac{e}{2S_n}\right)\Big/ D^{1/2}$$

$$X_2 = \frac{1}{2} - \left(\frac{w + e}{2S_c} - \frac{e}{2S_n}\right)\Big/ D^{1/2} \qquad (46)$$

The coefficients of the exponentials in compartment 2 are $\beta_2 X_1/(\beta_2 - \lambda_1)$ and $\beta_2 X_2/(\beta_2 - \lambda_2)$ which, upon introduction of the appropriate symbols and simplifying become $-e/S_n D^{1/2}$ and $+e/S_n D^{1/2}$ respectively. The final solutions are thus

$$\frac{a_c(t)}{a_c(0)} = \frac{1}{2}\left[1 + \left(\frac{1}{D^{1/2}}\right)\left(\frac{w + e}{S_c} - \frac{e}{S_n}\right)\right]e^{-\lambda_1 t}$$

$$+ \frac{1}{2}\left[1 - \left(\frac{1}{D^{1/2}}\right)\left(\frac{w + e}{S_c} - \frac{e}{S_n}\right)\right]e^{-\lambda_2 t}$$

$$\frac{a_n(t)}{a_c(0)} = \left(\frac{e}{S_n D^{1/2}}\right)(e^{-\lambda_2 t} - e^{-\lambda_1 t}) \qquad (47)$$

where $\lambda_{1,2}$ and D are defined in eq. 45.

Suppose that instead of washing out the initially labeled central compartment with non-labeled material, we have the system initially unlabeled but perfused with material of uniform specific activity $a(0)$. Here the same substitutions must be made, but compartment 1, the infinite compartment, is now the initially (and perpetually) labeled one. The general expressions are now given in eq. 42; thus, for the central compartment,

$$\frac{a_0(t)}{a_1(0)} = 1 + \left[\frac{B_1(\lambda_1 - \beta_2)e^{-\lambda_1 t}}{\lambda_1(\lambda_2 - \lambda_1)} \right] + \left[\frac{B_1(\lambda_2 - \beta_2)e^{-\lambda_2 t}}{\lambda_2(\lambda_1 - \lambda_2)} \right]$$

and for the lateral compartment:

$$\frac{a_2(t)}{a_1(0)} = 1 + \frac{B_1\beta_2}{\lambda_1 - \lambda_2}\left(\frac{e^{-\lambda_1 t}}{\lambda_1} - \frac{e^{-\lambda_2 t}}{\lambda_2} \right)$$

Inserting the appropriate notation of Fig. 15:

$$\frac{a_c(t)}{a(0)} = 1 - \frac{w}{S_c D^{1/2}}\left(\left\{ \frac{[(w+e)/S_c] - [e/S_n] + D^{1/2}}{[(w+e)/S_c] + [e/S_n] + D^{1/2}} \right\} e^{-\lambda_1 t} \right.$$

$$\left. - \left\{ \frac{[(w+e)/S_c] - [e/S_n] - D^{1/2}}{[(w+e)/S_c] + [e/S_n] - D^{1/2}} \right\} e^{-\lambda_2 t} \right)$$

$$\frac{a_n(t)}{a(0)} = 1 + \frac{2we}{S_c S_n D^{1/2}}\left(\left\{ \frac{e^{-\lambda_1 t}}{[e/S_n] + [(w+e)/S_c] + D^{1/2}} \right\} \right.$$

$$\left. - \left\{ \frac{e^{-\lambda_2 t}}{[e/S_n] + [(w+e)/S_c] - D^{1/2}} \right\} \right) \qquad (48)$$

The Superposition Principle

Up to this point we have considered only the system which contains a single initially labeled compartment. Actually, of course, the movement of label from each compartment is independent of movements from other compartments. It is a general property of linear systems that solutions of the equations can be formed by the superposition of several independent solutions. This characteristic provides the opportunity of expressing the specific activity in a compartment, when several compartments are initially labeled, as the sum of specific activities due to single initial-compartment labeling.

We first illustrate for the two-compartment system. Recalling, from

Chapter 1, the differential eqs. 1, which describe the system, we see that if compartment 1 alone is labeled, we have

$$a_1(t) = a_1(0)\left(\frac{S_1}{S} + \frac{S_2 e^{-\lambda t}}{S}\right)$$

$$a_2(t) = \frac{a_1(0)S_1}{S}(1 - e^{-\lambda t}) \qquad (49)$$

These expressions are acceptable as solutions, since they fulfill the necessary requirements. They not only satisfy the differential equations but also fit the initial condition that at zero time $a_1(t) = a_1(0)$ and $a_2(t) = 0$. Conversely, if a_2 is the initially labeled compartment,

$$a_1(t) = \frac{a_2(0)S_2}{S}(1 - e^{-\lambda t})$$

$$a_2(t) = a_2(0)\left(\frac{S_2}{S} + \frac{S_1 e^{-\lambda t}}{S}\right) \qquad (50)$$

These equations are obtained by reversing the numbering system for the compartments.

These solutions apply to their special cases, but, at times, it may be necessary to start an experiment after some redistribution of label has occurred. In this case each compartment initially has specific activity $a_1(0)$, $a_2(0)$, etc. From a well known theorem on linear differential equations (41), we know that, if the equations are soluble* and if any solution can be found which satisfies the equations and the specified initial conditions, any other solution is merely a redundant statement of the same mathematical relation. In the present case, where eqs. 49 and 50 are both solutions, application of the principle of superposition yields

$$a_1(t) = \frac{a_1(0)S_1 + a_2(0)S_2}{S} + S_2\left[\frac{a_1(0) - a_2(0)}{S}\right]e^{-\lambda t}$$

and
$$a_2(t) = \frac{a_1(0)S_1 + a_2(0)S_2}{S} + S_1\left[\frac{a_2(0) - a_1(0)}{S}\right]e^{-\lambda t} \qquad (51)$$

The substitution of these relations into the differential equations satisfies the original differential equations. Since, at $t = 0$,

$$a_1(t) = a_1(0)$$
and
$$a_2(t) = a_2(0)$$

* Criteria for solubility of linear differential equations will be found in standard reference works (41).

the initial conditions are also satisfied. Therefore these expressions are the desired solutions when both compartments are initially labeled.

We can illustrate this principle more generally for a mammillary system. Suppose that both compartments 0 and 1 are initially labeled. Then the numerator determinant of the Laplace transform for the central compartment may be expressed as follows:

$$\alpha_0 \Delta = \begin{vmatrix} a_0(0) & -B_1 & -B_2 & \cdots & -B_n \\ a_1(0) & (s+\beta_1) & 0 & \cdots & 0 \\ 0 & 0 & (s+\beta_2) & \cdots & 0 \\ \multicolumn{5}{c}{\cdots\cdots\cdots\cdots\cdots\cdots\cdots} \\ 0 & & & & (s+\beta_n) \end{vmatrix} \quad (52)$$

Expanding on minors by theorem 8, Chapter. 3,

$$\alpha_0 \Delta = a_0(0) \begin{vmatrix} s+\beta_1 & 0 & \cdots & 0 \\ 0 & (s+\beta_2) & \cdots & 0 \\ \multicolumn{4}{c}{\cdots\cdots\cdots\cdots\cdots} \\ 0 & 0 & \cdots & (s+\beta_n) \end{vmatrix}$$

$$+ a_1(0) \begin{vmatrix} B_1 & B_2 & \cdots & B_n \\ 0 & (s+\beta_2) & \cdots & 0 \\ \multicolumn{4}{c}{\cdots\cdots\cdots\cdots\cdots} \\ 0 & 0 & \cdots & (s+\beta_n) \end{vmatrix} \quad (53)$$

Thus the transform α_0 is the sum of the transforms for the two individual cases of α_0 where compartments zero and one, respectively, are initially labeled alone. Upon inversion the solution will be the superposition of the two respective solutions. This principle is obviously of considerable importance in the theory of tracer kinetics. An obvious extension to any system of initial labeling can be recognized and will apply to any system involving *linear differential equations*, whether or not that system can be treated by the Laplace transform method.

5

Other Steady-State and Quasi-Steady Systems, Approximation Methods

Besides mammillary systems, other systems may also be treated by the method of Laplace transforms. These include all systems which yield linear differential equations with constant coefficients, such as, for example, all closed systems which equilibrate by pure exchange. However, sometimes it is also possible to include systems in which "turnover" or "washout" of tracers occurs. Consider a catenary system through which material of zero or constant specific activity moves serially. Because as much substance enters a compartment as leaves it, the system will be in a steady state if the rate of transport through the system is constant from compartment to compartment and independent of time. The analysis also applies to the cyclic situation in which material leaving one end of a catenary system enters it once more at its beginning compartment. If a closed non-cyclic catenary system begins and ends with very large compartments, then the changes in these terminal compartments with time will be very slow, and we may consider them to be in a "quasi-steady state" (4).

Catenary Systems

The quasi-steady catenary system. Obtaining solutions of maximum generality will not add very much to what has been learned in previous chapters. However, information of general interest can be obtained from studies of the differential equations and their transforms.

We will consider a system such as the one in Fig. 16. In this illustration a constant amount of material, ϵ, is moving serially through the system in addition to exchange between compartments. The differential equations are as follows:

$$\frac{da_k}{dt} = \frac{1}{S_k} [a_{k-1}(\rho_{k-1} + \epsilon) + a_{k+1}\rho_k - a_k(\rho_{k-1} + \rho_k + \epsilon)] \tag{1}$$

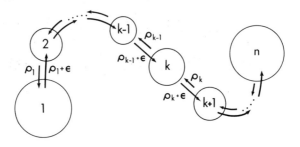

Fig. 16. A "quasi-steady" n-compartment catenary system showing hypothetical large first and last compartments. Transport between compartments occurs by exchange plus net turnover ϵ.

In obtaining the transforms we will use simplifying quantities:

$$d_j = [sS_j + (\rho_{j-1} + \rho_j + \epsilon)]$$
$$f_j = \rho_j + \epsilon$$

Taking the transform of eq. 1 and substituting the d's and f's, we obtain

$$-\alpha_{k-1}f_{k-1} + \alpha_k d_k - \alpha_{k+1}\rho_k = S_k a_k(0) \tag{2}$$

In a system with a first and last compartment, $f_0 = \rho_n = 0$, $d_n = sS_n + \rho_{n-1}$. The denominator determinant

$$\Delta(s) = \begin{vmatrix} d_1 & -\rho_1 & 0 & 0 & \cdot & \cdot & \cdot & 0 \\ -f_1 & d_2 & -\rho_2 & 0 & \cdot & \cdot & \cdot & 0 \\ 0 & -f_2 & d_3 & -\rho_3 & \cdot & \cdot & \cdot & 0 \\ \cdot & \cdot & \cdot & \cdot & \cdot & \cdot & \cdot & \cdot \\ 0 & 0 & \cdot & \cdot & \cdot & -f_{n-2} & d_{n-1} & -\rho_{n-1} \\ 0 & 0 & \cdot & \cdot & \cdot & 0 & -f_{n-1} & d_n \end{vmatrix} \tag{3}$$

will be a polynomial of degree n in s (through the d's which contain s). No general procedure suggests itself for obtaining the roots, although, if we define Δ_j as the determinant of rank j having the same elements as the first j rows and columns of Δ, we can show by theorem 8, Chapter 3, that

$$\Delta_j = d_j\Delta_{j-1} - \rho_{j-1}f_{j-1}\Delta_{j-2} \tag{4}$$

Methods for treating difference equations of this type usually require that the coefficients d_j and $\rho_{j-1}f_{j-1}$ be all alike for all values of j.

Because of the difficulties of the problem it will usually be expedient to expand the determinant in polynomial form and determine the roots numerically. In a fourth-rank determinant we find, by the expansion

$$\Delta_4 = d_1d_2d_3d_4 - d_1d_2f_3\rho_3 - d_1d_4\rho_2f_2 - f_1\rho_1d_3d_4 + f_1\rho_1\rho_3f_3 \tag{5}$$

that the expression is not symmetrical in the subscripts. Thus, even in this simple case, the roots will depend on the order of the subscripts and, therefore, on the order in which the compartments occur.

Pure exchange in a closed three-compartment catenary system. For pure exchange in a three-compartment catenary system with central compartment initially labeled, the solutions for central and peripheral compartments are given in eqs. 26 and 29 of Chapter 4. When a terminal compartment is labeled initially, we have the following expression:

$$\alpha_1 = \frac{\begin{vmatrix} a_1(0) & \dfrac{-\rho_{12}}{S_1} & 0 \\ 0 & s + \dfrac{\rho_{12}+\rho_{23}}{S_2} & \dfrac{-\rho_{23}}{S_2} \\ 0 & \dfrac{-\rho_{23}}{S_3} & s + \dfrac{\rho_{23}}{S_3} \end{vmatrix}}{\begin{vmatrix} s + \dfrac{\rho_{12}}{S_1} & \dfrac{-\rho_{12}}{S_1} & 0 \\ \dfrac{-\rho_{12}}{S_2} & s + \dfrac{\rho_{12}+\rho_{23}}{S_2} & \dfrac{-\rho_{23}}{S_2} \\ 0 & \dfrac{-\rho_{23}}{S_3} & s + \dfrac{\rho_{23}}{S_3} \end{vmatrix}}$$

$$= a_1(0)\left\{\frac{s^2 + [(\rho_{12}+\rho_{23})/S_2 + \rho_{23}/S_3]s + \rho_{23}\rho_{12}/S_2S_3}{s(s+\lambda_1)(s+\lambda_2)}\right\} \tag{6}$$

where $\quad \lambda_1\lambda_2 = (\rho_{12}\rho_{23})/(S_1S_2) + (\rho_{12}\rho_{23})/(S_1S_3) + (\rho_{12}\rho_{23})/(S_2S_3)$

and $\quad \lambda_1 + \lambda_2 = \rho_{12}(1/S_1 + 1/S_2) + \rho_{23}(1/S_2 + 1/S_3)$.

Here the compartments are considered as 1, 2, and 3, and thus the exchange rate between 1 and 2 is $\rho_{12} = \rho_{21}$, and between 2 and 3, $\rho_{23} = \rho_{32}$. This notation is different from that used for the mammillary system. Using that notation we may insert

$$\beta_1 = \rho_{12}/S_1 \qquad\qquad B_1 = \rho_{12}/S_2$$
$$\beta_2 = \rho_{23}/S_3 \qquad\qquad B_2 = \rho_{23}/S_2$$

into the above expressions for $\lambda_1\lambda_2$ and $\lambda_1 + \lambda_2$ and obtain the same results for λ_1 and λ_2 as given in eq. 23 in Chapter 4 for the mammillary system. The expression for α_1 inverts to

$$a_1(t) = a_1(0)\left[\frac{S_1}{S} + \frac{(\lambda_1{}^2\{[(\rho_{12} + \rho_{23})/S_2] + [\rho_{23}/S_3]\}\lambda_1 + \{\rho_{23}\rho_{12}/S_2 S_3\})e^{-\lambda_1 t}}{\lambda_1(\lambda_1 - \lambda_2)}\right.$$
$$\left. + \frac{(\lambda_2{}^2\{[(\rho_{12} + \rho_{23})/S_2] + [\rho_{23}/S_3]\}\lambda_2 + \{\rho_{23}\rho_{12}/S_2 S_3\})e^{-\lambda_2 t}}{\lambda_2(\lambda_2 - \lambda_1)}\right]$$

$$(7)$$

For the central compartment

$$\alpha_2 = \frac{a_1(0)[(\rho_{12}s/S_2) + (\rho_{12}\rho_{23}/S_2 S_3)]}{s(s + \lambda_1)(s + \lambda_2)}$$

yielding

$$a_2(t) = a_1(0)\left\{\frac{S_1}{S} + \frac{[\rho_{12}/S_2][(\rho_{23}/S_3) - \lambda_1]e^{-\lambda_1 t}}{\lambda_1(\lambda_1 - \lambda_2)}\right.$$
$$\left. + \frac{[\rho_{12}/S_2][(\rho_{23}/S_3) - \lambda_2]e^{-\lambda_2 t}}{\lambda_2(\lambda_2 - \lambda_1)}\right\} \qquad (8)$$

For the end compartment

$$\alpha_3 = a_1(0)\,\frac{\rho_{12}\rho_{23}/S_2 S_3}{s(s + \lambda_1)(s + \lambda_2)}$$

yielding
$$a_3(t) = \frac{a_1(0)S_1}{S}\left[1 + \left(\frac{\lambda_2 e^{-\lambda_1 t} - \lambda_1 e^{-\lambda_2 t}}{\lambda_1 - \lambda_2}\right)\right] \qquad (9)$$

It will be noted that, since there are as many exponential constants as there are exchange rates in systems of this sort, it should be possible to determine the kinetics of the system from sufficiently precise observations in only one compartment. This does not violate the principles enunciated in Chapter 3 because the solution depends on the assumption of equal and opposite constant rates between compartments. Such assumptions are typical of information which must be advanced from sources other than tracer analyses in order that meaningful information may be obtained.

Simple turnover in catenary systems. One situation which has been of interest to isotope workers is simple "washout" or "turnover" through a catenary system without exchange. Here, if we insert zero for all ρ's, the denominator determinant becomes

$$\Delta = d_1 d_2 \cdots d_n = \left[\left(s + \frac{\epsilon}{S_1} \right) \left(s + \frac{\epsilon}{S_2} \right) \cdots \left(s + \frac{\epsilon}{S_n} \right) \right] (S_1 S_2 \cdots S_n) \quad (10)$$

If the ith compartment is labeled and observations are made in the jth, the i will be the row index and the j the column index in the determinant and we have

$$\alpha_j \Delta = \begin{vmatrix} d_1 & 0 & 0 & \cdots & 0 & 0 & \cdots & 0 & \cdots & 0 & 0 \\ -\epsilon & d_2 & 0 & \cdots & 0 & 0 & \cdots & 0 & \cdots & 0 & 0 \\ 0 & -\epsilon & d_3 & \cdots & 0 & 0 & \cdots & 0 & \cdots & 0 & 0 \\ \cdots & & & & & & & & & & \\ 0 & 0 & 0 & \cdots & 0 & 0 & \cdots & 0 & \cdots & 0 & 0 \\ 0 & 0 & 0 & \cdots & 0 & d_{j+1} & \cdots & 0 & \cdots & 0 & 0 \\ \cdots & & & & & & & & & & \\ 0 & 0 & 0 & \cdots & S_i a_i(0) & 0 & \cdots & d_i & \cdots & 0 & 0 \\ \cdots & & & & & & & & & & \\ 0 & 0 & 0 & \cdots & 0 & 0 & \cdots & 0 & \cdots & d_{n-1} & 0 \\ 0 & 0 & 0 & \cdots & 0 & 0 & \cdots & 0 & \cdots & -\epsilon & d_n \end{vmatrix}$$

$$(11)$$

If $j < i$, the $a_i(0)$ term will lie to the left of the diagonal yielding a determinant of the form illustrated in theorem 4, Chapter 3. However, because of a zero on the diagonal, the determinant will equal zero. This is the consequence of observing a compartment in the chain above the point of labeling.

If we label and observe in the same compartment, that is if $j = i$, then

$$\alpha_j = \frac{a(0)(s + \epsilon/S_1) \cdots (s + \epsilon/S_{j-1})(s + \epsilon/S_{j+1}) \cdots (s + \epsilon/S_n)}{(s + \epsilon/S_1) \cdots (s + \epsilon/S_n)} = \frac{a_j(0)}{s + \epsilon/S_j}$$

$$a(t) = a_j(0) e^{-\epsilon t/S_j} \quad (12)$$

From the point of view of physics, it may be appreciated that compartments below the point of observation will not affect the results at the observation point.

When $a_i(0)$ is to the right of the diagonal, we first form the product of $a_i(0)$ and its minor. The minor in turn may be expanded by the elements of the first column and progressively reduced in rank yielding, finally,

$$\alpha_j \Delta = S_i a_i(0)\, d_1\, d_2 \cdots d_{i-1} \epsilon^{j-i}\, d_{j+1} \cdots d_n \tag{13}$$

Thus the final transform is

$$\alpha_j = S_i a_i(0) \frac{\epsilon^{j-i}}{(sS_i + \epsilon)(sS_{i+1} + \epsilon)\cdots(sS_j + \epsilon)}$$

$$= a_i(0) \frac{\lambda_{i+1}\cdots\lambda_j}{(s + \lambda_i)(s + \lambda_{i+1})\cdots(s + \lambda_j)} \tag{14}$$

From the form of this expression it is seen that the result depends only on the compartments between and including the initially labeled one and the one under observation. For this reason we will consider the labeled one as number 1 and the observational as the nth. In these terms the final solution is

$$a_n(t) = a_1(0)\lambda_2\lambda_3 \cdots \lambda_n \sum_{k=1}^{k=n} \frac{e^{-\lambda_k t}}{\prod_{\substack{r=1 \\ k \neq r}}^{r=n} (\lambda_r - \lambda_k)} \tag{15}$$

where $\lambda_k = \epsilon/S_k$.

In simple washout of a non-cyclic catenary system it is of interest to note that the solution for compartment j does not depend on the order of the compartments between it and the initially labeled one. We may interchange the intermediate compartments without affecting the terminal compartment.

Cyclic Systems

When the initial and terminal compartments of a catenary system are brought into communication, it becomes a cyclic system (Fig. 17). We

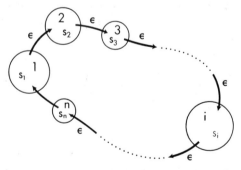

Fig. 17. A cyclic system in which material is transported by simple turnover at a rate ϵ.

will consider simple turnover without exchange. A number of interesting properties are found which are different from those of a linear system. The system is also of interest in connection with lumped analogs for simulation of circulatory mixing which will be discussed in Chapter 9. In the analysis which follows we will adopt the simplifying procedure of dividing the ith row of the numerator and denominator by S_i. For simple turnover without exchange where the ith compartment is initially labeled, the transform for compartment j becomes

$$\alpha_j = \frac{\begin{vmatrix} (s + \epsilon/S_1) & 0 & \cdots & 0 & 0 & \cdots & 0 & -\epsilon/S_1 \\ -\epsilon/S_2 & (s + \epsilon/S_2) & \cdots & 0 & 0 & \cdots & 0 & 0 \\ \cdots\cdots\cdots\cdots\cdots\cdots\cdots\cdots\cdots\cdots\cdots\cdots\cdots\cdots\cdots \\ 0 & 0 & \cdots & (s + \epsilon/S_{j-1}) & 0 & \cdots & 0 & 0 \\ \cdots\cdots\cdots\cdots\cdots\cdots\cdots\cdots\cdots\cdots\cdots\cdots\cdots\cdots\cdots \\ 0 & 0 & \cdots & 0 & a_i(0) & \cdots & 0 & 0 \\ \cdots\cdots\cdots\cdots\cdots\cdots\cdots\cdots\cdots\cdots\cdots\cdots\cdots\cdots\cdots \\ 0 & 0 & \cdots & 0 & 0 & \cdots & -\epsilon/S_n & (s + \epsilon/S_n) \end{vmatrix}}{\begin{vmatrix} (s + \epsilon/S_1) & 0 & & \cdots & & 0 & -\epsilon/S_1 \\ -\epsilon/S_2 & (s + \epsilon/S_2) & & \cdots & & 0 & 0 \\ \cdots\cdots\cdots\cdots\cdots\cdots\cdots\cdots\cdots\cdots\cdots\cdots\cdots \\ 0 & 0 & & \cdots & & -\epsilon/S_n & (s + \epsilon/S_n) \end{vmatrix}} \tag{16}$$

At this point we notice one of the important differences which distinguish a cyclic system from a linear one. As a result of the term in the upper right corner of the determinant the roots now become complex numbers, and the solutions contain an oscillatory component. The denominator, expanded by minors using the elements of the last row (theorem 8, Chapter 3), yields

$$\Delta = \left(s + \frac{\epsilon}{S_1}\right)\left(s + \frac{\epsilon}{S_2}\right) \cdots \left(s + \frac{\epsilon}{S_n}\right) - \frac{\epsilon^n}{S_1 S_2 \cdots S_n} \tag{17}$$

The numerator will no longer necessarily be zero if $j < i$. From the point of view of physics, this statement implies that a compartment which is "downstream" for one cycle of indicator around the system is "upstream" for the next. As before, the case where $j \geq i$ will cover all possibilities. The numerator is the same as in the linear case because in the numerator the minor of the upper right hand corner term will always be zero. Therefore, we have

$$\alpha_j = a_i(0) \frac{(s + \beta_1) \cdots (s + \beta_{i-1})(-\beta_i) \cdots (-\beta_j)(s + \beta_{j+1}) \cdots (s + \beta_n)}{\Delta}$$

for $j \neq i$, and

$$\alpha_i = a_i(0) \frac{(s + \beta_1) \cdots (s + \beta_{i-1})(s + \beta_{i+1}) \cdots (s + \beta_n)}{\Delta} \tag{18}$$

for $j = i$, where $\beta_i = \epsilon/S_i$.

On examining Δ, we note that the term $\epsilon^n/S_1 S_2 \cdots S_n$ just cancels the constant term of the product of $(s + \epsilon/S_n)$ factors. Therefore the denominator of α_j is a product of s and a polynomial of degree $n-1$ with roots $-\lambda_1, -\lambda_2, \cdots, \lambda_{n-1}$. The presence of s as a lone factor means that one root is zero; thus the inverse contains a constant term. This

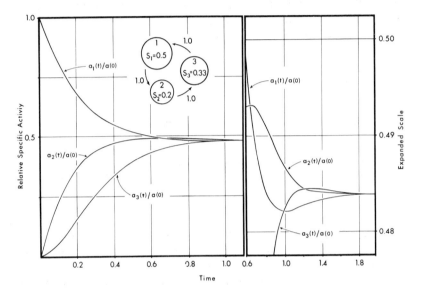

Fig. 18. Specific-activity changes in a three-compartment cyclic system. The compartmental system and turnover rate are shown in the left-hand figure. The right-hand figure with the expanded vertical scale shows the details of the interlacing of the curves.

term will be the specific activity at infinite time and it will not approach zero. From physical considerations it is seen that the ultimate specific activity will be $(a_i(0)S_i)/S$ as in other situations. The individual exponential constants in the non-cyclic serial case depend only on the properties of individual compartments, but in the cyclic system this simplification is lacking.

Finally then

$$a_j(t) = a_i(0)\left(\frac{S_i}{S} + \sum_{k=1}^{k=n} Y_k e^{-\lambda_k t}\right)$$

where

$$Y_k = \frac{(\beta_1 - \lambda_k) \cdots (\beta_{i-1} - \lambda_k)(-\beta_i) \cdots (-\beta_j)(\beta_{j+1} - \lambda_k) \cdots (\beta_n - \lambda_k)}{(-\lambda_k)(\lambda_1 - \lambda_k) \cdots (\lambda_{k-1} - \lambda_k)(\lambda_{k+1} - \lambda_k) \cdots (\lambda_n - \lambda_k)}$$

$$(19)$$

A plot of a typical set of a's for a three-compartment cyclic system is given in Fig. 18. It is of interest to note that at every point where $a_{k-1}(t)$ and $a_k(t)$ intersect, $a_k(t)$ has a maximum or a minimum value. This fact is readily seen from the differential equation for this pair of members.

$$S_k \frac{da_k}{dt} = \epsilon(a_{k-1} - a_k) \qquad (20)$$

At the point of intersection of a_{k-1} and a_k their difference is zero. Therefore $da_k/dt = 0$ and as a result a_k has a maximum or minimum. Zilversmit and his associates have pointed out that, for non-cyclic systems, this relation is a necessary consequence of the fact that compartment $k-1$ is an immediate precursor of compartment k (42). In the more general cyclic system, every compartment is a precursor of every other although not always an immediate one. Of course a linear system might be considered as a portion of an infinite cyclic system, but, in a very large system, the minima which may occur will be so flat and late in time that they will be of no practical importance. Therefore only the first maximum in compartment k will be recognized.

Approximation Methods

The use of approximate models in physics has been one of the most powerful tools of the theoretician. On the other hand, the biologist has often regarded this approach with suspicion. The value of well-chosen approximations lies in the ability of the approximator to temporarily exclude non-essential details from his thinking. Thus, for example, the general precise theory of refraction of light rays at a spherical surface tells us that a lens cannot converge rays to a common intersection. This true fact is beside the point in most problems of practical optics. The small-angle approximation method permits us to develop a useful theory of lens systems which can then be refined by using a second approximation that is closer to the facts and yields the theory of optical aberrations. In tracer kinetics it may often be possible to see the problem more clearly in the more complicated systems if we use approximation methods.

Systems with widely differing rates.   Occasionally in tracer kinetic expressions considerable simplification can be achieved if certain terms can be neglected or replaced by other nearly equal quantities which are more mathematically tractable. On the one hand, the art of approximation is hazardous, and we must be cautious in its use. On the other, little is to be gained in encumbering a mathematical expression with many terms if the fractional amount that is thereby added or subtracted to or from the result is negligible.

An illustrative example of the treachery involved in approximation is seen in a two-compartment mammillary system where one exchange rate is much greater than the other. Suppose that the fast compartment is number 1. Then, in the discriminant (eq. 24, in Chapter 4), we have $\beta_1 \gg \beta_2$, $B_1 \gg B_2$. Apparently we may insert $\beta_2 = B_2 = 0$ with sufficient precision into the equation. Thus the expression for D approaches

$$\beta_1{}^2 + B_1{}^2 + 2\beta_1 B_1 = (\beta_1 + B_1)^2$$

the λ's then become

$$\lambda_1 = \beta_1 + B_1$$
$$\lambda_2 = \frac{\beta_1 + B_1 - \beta_1 - B_1}{2} = 0$$

which are the exponential constants for a two-compartment system. Obviously the art of approximation has gone too far in this case, since λ_2 should not vanish entirely.

At this point, however, we note that D will be a perfect square if we replace $(B_1 + B_2)^2$ by $(B_1 - B_2)^2$. If $B_1 \gg B_2$, this substitution is permissible and yields

$$D^{1/2} = \beta_1 - \beta_2 + B_1 - B_2$$
$$\lambda_1 = B_1 + \beta_1 \tag{21}$$

In the above, λ_1 is the same as before but D is different. In this special case the exponential constant for the fast compartment depends only on properties of that compartment. In terms of physics, the slow compartment is changing too slowly to have appreciable influence. In the case of λ_2, however, this quantity is a small difference between large quantities whose small variations can produce serious effects. Thus we proceed instead to invoke the relation that the product of the roots is equal to the constant term in the numerator of $L(s)$. This term is readily obtained by inserting $s = 0$ into $L(s)$. Thus

$$L(0) = \left(1 + \frac{B_1}{\beta_1} + \frac{B_2}{\beta_2}\right)$$
$$= \frac{\beta_1 \beta_2 + \beta_2 B_1 + \beta_1 B_2}{\beta_1 \beta_2}$$

therefore

$$\lambda_1 \lambda_2 = \beta_1 \beta_2 + \beta_2 B_1 + \beta_1 B_2$$
$$\lambda_2 = \frac{\beta_1 \beta_2 + \beta_2 B_1 + \beta_1 B_2}{B_1 + \beta_1}$$
$$= \beta_2 + \frac{\beta_1 B_2}{B_1 + \beta_1} \tag{22}$$

This procedure provides considerable simplification in obtaining the λ's, an operation which usually represents a great portion of the total labor in problems of this sort. In general, wherever approximation methods suggest themselves, they may well be considered. In considering them, we must keep in mind which terms are the small terms and in relation to what they are considered small.

The physical significance of the above expression is clear for λ_1, since this expression describes exchange between the central and fast peripheral compartments. For λ_2 we can write

$$\lambda_2 = \beta_2 + \frac{B_2\rho_1}{S_1(\rho_1/S_0 + \rho_1/S_1)} = \beta_2 + \frac{\rho_2}{S_0 + S_1} \tag{23}$$

which is the exponential constant for equilibration between compartment 2 and combined compartments 0 and 1.

Numerical reduction of expressions to particular cases. The material up to this point has emphasized general solutions and their reduction to various algebraic expressions to fit particular cases. There is an advantage in this approach when certain general conclusions are to be drawn. For example, by the general approach one may predict that in one type of system the final specific activity will always be zero, whereas in another it will not. Sometimes, predictions on a mathematical basis will also yield information not immediately obvious from purely physical considerations alone.

Those using the tracer method, however, will often wish to know what the actual behavior of some particular system might be. We will therefore illustrate the procedure which might be adopted in working out numerical results for a particular system—for example, the system of the previous section. The experiment which is to be compared with these theoretical results is the equilibration of 3 cc of canine blood (43) in vitro with plasma containing K^{42}. In some experiments the "buffy coat" fraction is removed and in others it is not. In this way by separating cells from plasma periodically and analyzing for K^{42}, specific activity curves can be constructed for the plasma (compartment 0), and the mature erythrocytes (compartment 2). By difference one obtains data for the "buffy coat" compartment (number 1). To set up the theoretical specific activity curves we may tentatively use the following values (amounts in microequivalents—μeq—of K and rates in microequivalents of K per day):

$$S_0 = \quad 6 \ \mu\text{eq} \qquad S_1 = 3 \ \mu\text{eq} \qquad S_2 = 6 \ \mu\text{eq} \qquad S = 15 \ \mu\text{eq}$$

$$\rho_1 = 100 \ \mu\text{eq per day} \qquad\qquad \rho_2 = 1.5 \ \mu\text{eq per day}$$

These values yield for the β's and B's (in reciprocal days)

$$\beta_1 = 33.3 \qquad B_1 = 16.7$$

$$\beta_2 = 0.25 \qquad B_2 = 0.25$$

It is seen that the requirement $B_1 \gg B_2$ is reasonably well satisfied. Thus $(B_1 + B_2)^2 = 287.3$ whereas $(B_1 - B_2)^2 = 270.6$, a difference of about 6 per cent, which will be proportionately reduced in the final evaluation of D and of the λ's. From eqs. 21 we have

$$D = 2450$$

and

$$\lambda_1 = 50$$

From eq. 22

$$\lambda_2 = 0.25 + \frac{33.3 \times 0.25}{50} = 0.416$$

If we insert these values of λ_1 and λ_2 in eqs. 26 and 29 in Chapter 4 respectively,

$$a_0(t) = a_0(0)(0.4 + 0.335e^{-50t} + 0.265e^{-0.416t})$$

and

$$a_1(t) = a_0(0)(0.4 - 0.67e^{-50t} + 0.27e^{-0.416t})$$

A similar expression can readily be written for $a_2(t)$. For comparison purposes, if we had not used the approximation, the result would be

$$D = 2466.95$$

yielding

$$\lambda_1 = \frac{50.5 + 49.67}{2} = 50.08$$

and

$$\lambda_2 = 0.42$$

These values of λ are scarcely different from those obtained by the approximation procedure.

The specific activities are always proportional to the initial value $a_0(0)$. Since compartment sizes remain constant, the ratio of $a_0(t)/a_0(0)$ is the same as the ratio of the activities of compartment zero. In any given instance the activity of this compartment is thus obtained by multiplying the initial activity by the ratio of $a_0(t)/a_0(0)$. This ratio is independent of initial activity and may thus be plotted for all experiments which might be conducted in that particular system (Fig. 19). For the peripheral compartment—say, number 1—the activity at any time will be obtained by multiplying the initial activity by the ratio S_1/S_0 in addition to $a_0(t)/a_0(0)$.

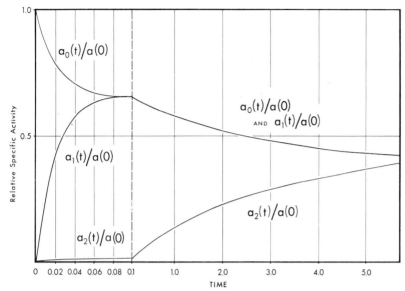

Fig. 19. Specific activity changes in a three-compartment mammillary system in which the exchange rates with the peripheral compartments differ greatly. Note time scale change at $t = 0.1$.

Perturbation methods. If we consider for the moment the kth compartment of a multicompartment system we may write the equation for it as

$$S_k \frac{da_k}{dt} = \sum_j \rho_{kj} a_j - a_k \sum_j \rho_{jk}$$

or

$$S_k \frac{da_k}{dt} + a_k \sum_j \rho_{jk} = \sum_j \rho_{kj} a_j \qquad (24)$$

If the $a_j(t)$ are all known as a function of time, then this differential equation may be solved by relatively simple methods. Suppose that, by removal of a_k from the system, the solution could be easily determined for the rest of the system. Here, a rather simple extension could be performed to include a_k. The difficulty is that, as soon as a_k is added, not only does the rest of the system affect a_k, but a_k affects the rest of the system as well.

However, there are frequent cases where the effect of a_k is small—for example, where a_k is a small compartment in a large system. The approach, then, is to solve the system without a_k, insert the solution in eq. 24 and solve for a_k. This method is known as the method of perturbations. The first step which we have described is the first approximation to the solution for a_k. Often this will give a sufficiently precise answer. However, if more precision is desired, we can determine the modification which a_k produces in the system and proceed to determine from this a better solution for a_k. Proceeding by steps, we can thus obtain a second approximation and higher approximations. Usually, however, the labor does not justify going beyond the first.

The method will be particularly advantageous in a system where compartment k makes only one connection with the residual system, say at compartment r. Here we may write the Laplace transform for compartment k:

$$\alpha_k(sS_k + \rho_{rk}) = \rho_{kr}\alpha_r$$

Thus

$$\alpha_k = \left(\frac{\rho_{kr}}{sS_k + \rho_{rk}}\right)\alpha_r \qquad (25)$$

and the transform for compartment k is obtained as the product of a constant factor and of the transform for k alone times the transform for r. This general rule for situations of this sort will often prove to be of considerable use.

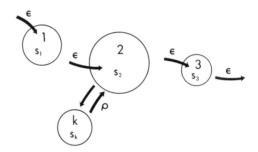

Fig. 20. The use of perturbation methods. Compartment k is considered to have negligible effect on compartments 1, 2, and 3. Compartment 2 "drives" compartment k just as it drives compartment 3.

We can illustrate in the case of the system in Fig. 20. This four-compartment mammillo-catenary system would be cumbersome to solve by standard methods. Using the approximation method from eq. 25,

$$\alpha_k = \left(\frac{\rho}{sS_k + \rho}\right)\alpha_2$$

From eq. 14 we have

$$\alpha_2 = \frac{S_1 a_1(0)\epsilon}{(sS_1 + \epsilon)(sS_2 + \epsilon)}$$

yielding

$$\alpha_k = \frac{S_1 a_1(0)\epsilon\rho}{(sS_1 + \epsilon)(sS_2 + \epsilon)(sS_k + \rho)} \qquad (26)$$

The transform in eq. 26 is similar in form to that for the third compartment of a catenary system with turnover. The reason for this is that we assume in this approximation that the compartment does not affect compartment 2 with which it communicates. This is an expression of the fact that the activity returning to 2 from k in the exchange process is negligible. In simple turnover the preceding compartment affects the one which follows, but is not affected by it. The validity of the method in this case rests on the smallness of compartment k relative to compartment 2 since this will determine the amount of its effect. The final approximate solution will be analogous to eq. 15, namely

$$a_k(t) = \lambda_2 \rho a_1(0)\left[\frac{e^{-\lambda_1 t}}{(\lambda_2 - \lambda_1)(\lambda_k - \lambda_1)} + \frac{e^{-\lambda_2 t}}{(\lambda_k - \lambda_2)(\lambda_1 - \lambda_2)}\right.$$
$$\left. + \frac{e^{-\lambda_k t}}{(\lambda_1 - \lambda_k)(\lambda_2 - \lambda_k)}\right] \quad (27)$$

where $\lambda_1 = \epsilon/S_1$, $\lambda_2 = \epsilon/S_2$, and $\lambda_k = \rho/S_k$

Of course a higher approximation step would be to take into account the effect of k on compartment 2, as though k were precisely represented by eq. 27. From this a new expression for $a_k(t)$ could be derived.

The convolution theorem, driving and driven compartments. In equations such as 26, where Laplace transforms occur as products of individual transforms, whose inverses may be readily recognized, a convenient theorem can often be used to simplify the inversion process.

Given

$$\mathscr{L}f_1(t) = \Phi_1(s), \ \mathscr{L}f_2(t) = \Phi_2(s)$$

then

$$\mathscr{L}\left[\int_0^t f_1(t - T)f_2(T)\,dT\right] = \Phi_1(s)\Phi_2(s) \quad (28)$$

The integral on the left is called the "convolution" of f_1 and f_2 and will be further discussed in Chapter 9. It is encountered in the literature on electric circuit analysis, heat flow and other branches of mathematical physics.

Suppose we have a system such as the one in Fig. 15. We have discussed its behavior already for two typical situations from one point of view but the use of the convolution theorem will also be instructive here. We rewrite the differential equations for compartments c and n,

$$\frac{da_c}{dt} = \frac{e}{S_c}(a_n - a_c) + \frac{wF_\infty(t)}{S_c} - \frac{wa_c}{S_c}$$

$$\frac{da_n}{dt} = \frac{e}{S_n}(a_c - a_n) \quad (29)$$

where $F_\infty(t)$ is now the specific activity of material entering from the "infinite compartment." Suppose that the system is initially unlabeled. We may write the transforms as

$$\alpha_c(sS_c + e + w) - e\alpha_n = w\Phi_\infty(s)$$
$$-e\alpha_c + (sS_n + e)\alpha_n = 0 \tag{30}$$

where $\Phi_\infty(s)$ is the transform of $F_\infty(t)$. We recognize that the equations are identical with the equations for the system where compartment c initially contains $a(0)$ except that $w\Phi_\infty(s)$ takes the place of the factor $S_c a(0)$. This analysis can be generalized to any system whatever having a single initially labeled compartment. In every case where $a(0)$ was initially a factor in the transform for a particular compartment we now have $w\Phi_\infty(s)/S_c$ as a factor. Suppose in the case of compartment r we originally had

$$\alpha_r = x(s)a(0)$$

the generalization yields

$$\alpha_r = \frac{x(s)w\Phi_\infty(s)}{S_c}$$

If originally the response of compartment k was $a_k(t)$ by the convolution theorem it will now be

$$a_k(t) = \frac{w}{S_c} \int_0^t a_k(t - T)F_\infty(T)\, dT \tag{31}$$

We may regard $F_\infty(t)$ as a "forcing function" to which the system responds. When only compartment k was initially labeled, this could only have been achieved if label was introduced suddenly from the infinite compartment during such a brief time that no label penetrated into other compartments. This would be achieved if $wF_\infty(t)/S_c$ were zero and were "turned on" for a very brief moment and then turned off. This impulse function representation of $wF_\infty(t)/S_c$ would then have the Laplace transform $a(0)$.

This approach can be extended to predict the system response to various types of $F_\infty(t)$. For the case where $F_\infty(t)$ is initially zero, being switched on at $t = 0$ we use the transform for a "unit step." This can best be determined as the limit of $e^{-\lambda t}$ for $\lambda \to 0$. From Table 1, we see that the transform is $1/s$. The procedure to follow in this case will depend upon convenience. If the transform of the system is given, it may be possible to multiply by $1/s$ and invert by searching tables of transforms. In other cases one must use eq. 28. If $F_\infty(\lambda)$ is a unit step, then by elementary calculus

$$a_k(t) = \int_0^t a_k(T)\, dT$$

where $a_k(T)$ is the response to an "impulse" of label.

We may illustrate the preceding considerations in a one-compartment system containing S washed out at a rate w with tracer of specific activity $F_\infty(t)$. The system transform for the case of initial labeling is

$$\alpha(s) = \frac{a(0)}{s + w/S}$$

yielding $\qquad\qquad a(t) = a(0)e^{-wt/S}$

In the general case we have

$$\alpha(s) = \frac{w\Phi_\infty(s)}{S} \times \frac{1}{s + w/S} \tag{32}$$

with inverse

$$a(t) = \frac{w}{S} \int_0^t e^{-(w/S)(t-T)}F_\infty(T)\,dT \tag{33}$$

For a unit step

$$\alpha(s) = \frac{w}{Ss(s + w/S)} \tag{34}$$

yielding

$$a(t) = \frac{w}{S}\int_0^t e^{-wT/S}\,dT$$

$$= 1 - e^{-wt/S} \tag{35}$$

Of course in the present case $F_\infty(t)$ may be any arbitrary function of time. When it represents the specific activity of one compartment of a multicompartment system "driving" a small compartment, we may recognize the situation as being a general situation of which eq. 25 is a special case. Where it is inconvenient to work solely in transform space, we may obtain the specific activity in compartment k by a direct convolution integration. Of course the principle is not restricted to a single compartment and can be extended to include one multicompartment system "driven" by another. The practical situations, however, will usually involve the simpler systems.

Practical procedures for solving multicompartment systems. Of course a few simpler systems may be recognized as special instances of some system whose solution is already known. By inserting numerical values into the formulas for these systems and using the concept of infinite compartments, we may sometimes be able to obtain a desired result. In other cases, we may only require the specific activity for one or two compartments in a more complicated system. Here it may often be possible to lump several compartments together and consider them as one average compartment. If the specific activity of one of the subcompartments is then required, it may be possible to obtain a good approximation by considering that subcompartment as being "driven" by the remainder of the system. If all other methods fail and an analytical solution is required we must usually proceed directly by the method of Laplace transforms. Sometimes, the principle of superposition may be used to combine solutions for one set of initial conditions to obtain results for another set.

6

Analog Simulation of Compartmental Tracer Experiments

When the variables which describe the physical behavior of one system behave analogously to those which describe another, the two systems are said to be analogs of each other. Analogous behavior is usually recognized when the equations for both systems have the same form. The convenience of an analog is clear if it is more readily investigated or manipulated than the system which it is to simulate. If there is one conclusion to be drawn from the preceding chapters, it is that the kinetics of tracer experiments may at times involve equations of considerable complexity and their solution may be quite laborious. The simple problem, for example, of substituting numerical data into equations for even relatively simple situations and solving for exponential constants can be quite tedious. In systems of more than three compartments it will often be necessary to obtain roots of cubic or quartic equations or equations of even higher degree.

Even if general formulas can be written out for solutions (which is often not possible), the labor of inserting numbers and performing the arithmetic calculations may be considerable. It is therefore clear that the simulation of tracer kinetics by physical analogs has considerable merit. Because of the flexibility of some types of analogs it is often possible to simulate several possible models almost as readily as one. This is the best safeguard against making the wrong unique choice among various competing models. The simulating systems which have received most attention have been hydrodynamic analogs (10, 44, 45, 46, 47) and electrical analogs (4, 48, 49, 50, 51).

Non-Electrical Types of Analogs

Quite a wide variety of analogs to tracer experiments may be imagined. There is a basic similarity between the equations of tracer kinetics and the classical equation which describes general diffusion problems including heat conduction. More about this subject will be found in Chapter 8. In the search for analogs (if we were to select a possibly extreme situation), we might conceptually simulate tracer experiments by following the movement of compressible fluids in beds of sand or among rigid reservoirs, since these systems follow similar equations. We can also imagine a thermal analogy based on the observation of heat flow between metal blocks separated by material of low thermal conductivity. The practical problem is to select from these various analog systems those which can be used most conveniently.

One practical method of simulation is through the use of hydrodynamic systems. Water is circulated between stirred vessels, and dye is followed after injection. This method has the disadvantage of being slow and of requiring the handling of liquid samples. If certain dyes are used, some care as to the effect of pH changes and other chemical effects may be necessary. However, many laboratories are equipped for precise colorimetry. In addition, the equipment for such simulations can often be readily found in a conventional research laboratory. These analogs are also advantageous when used in lecture demonstrations to students, since we can easily see what is happening in the system. They have a scientific advantage in that they permit us to study systems which are not in a steady state almost as readily as those which are.

Figure 21 shows a typical hydrodynamic analog system. Water is moved from vessel to vessel by pumps and by gravity. Each vessel represents a tracer compartment whose size is represented by the volume of water in the vessel. Uniform mixing is insured by the use of stirrers in the compartments. Transport rates are simulated by the quantitated rates of water movement between compartments. Dye is injected into one of the compartments at zero time. Its initial concentration is analogous to initial specific activity. At various times small samples are obtained from the compartments, and the dye concentration is determined.

The dye concentration is analogous to $a(t)$, and the equations describing such systems are readily obtained. Let the volume of compartment i be $\{V_i\}$. The concentration $[C_i]$ of dye in the compartment may be multiplied by $\{V_i\}$ to give the amount of dye in the compartment. The rate of change of this quantity is the difference between the rate at which dye enters and leaves. These rates are obtained by summation over the various other compartments. In general, the rate at which dye moves from j to i is equal to the rate of water movement r_{ij} times the concentration of dye in the originating compartment $[C_j]$. Thus we can obtain the equation

$$\frac{d([C_i]\{V_i\})}{dt} = \sum_j r_{ij}[C_j] - \sum_j r_{ji}[C_i] \tag{1}$$

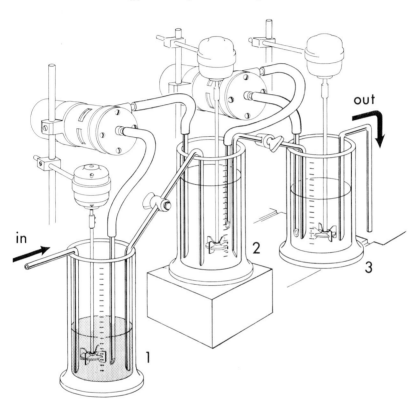

Fig. 21. A hydrodynamic analog system. Fluid moves between stirred reservoirs, being conveyed by pumps, gravity, or pressure. Reservoir 1 is initially labeled, and the movement of label is followed as a function of time.

Comparing this with eq. 2 in Chapter 3, and recalling that there, $R_i = a_i S_i$ (R_i being *tracer*), we find that the analogy between the corresponding quantities is readily established. Here we note that both $[C_i]$ and $\{V_i\}$ are included in the differentiation since either quantity may conceivably vary. For the steady-state system, $\{V_i\}$ may be factored out and placed before the derivative.

Resistance-Capacitance Networks as Electrical Analogs

The flexibility, cleanliness, and convenience of electrical analogs make them highly advantageous in the simulation of steady-state systems. Here they are the method of choice for routine use. When used as quantitative devices, they may be termed "analog computers," and as such they have been used for many years by physicists and engineers for solving differential equations. Considerable discussion will therefore be given to this type of simulation. Many electrical analogs include electronic amplifiers.

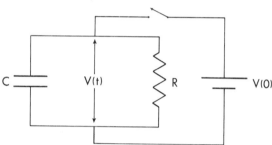

Fig. 22. A simple resistance-capacitance circuit. Battery voltage $V(0)$ is applied through the switch. This is opened at zero time, and voltage $V(t)$ across R and C is followed as a function of time.

Of course electrical analogs which contain amplifiers have the disadvantage that the electrical events within the system are not readily apparent to those unfamiliar with electronics. Nevertheless, since only certain types of equations will be solved, it is not necessary to make an exhaustive study of electronic analog computation methods to gain special facility with the methods applicable to tracer kinetic problems. Just as it is not essential to understand the internal combustion engine to drive an automobile, much of the electronic basis of the analog systems need not be explored. Of course, experience has shown that those who try these methods usually become familiar with certain of the basic electronic principles and develop an interest in this field. In the present outline we will avoid an exhaustive inquiry into general electronic fundamentals which can be found elsewhere in the literature.

The key to tracer simulation lies in the electrical behavior of systems of capacitances and resistors. Let us consider for the moment a condenser which has been charged by a battery and then allowed to discharge through a resistance (Fig. 22). After the battery is charged to $V(0)$, if the switch is opened, the current through the resistance is determined by the voltage which the condenser maintains across it. At the same time the current represents the rate of discharge of the condenser. As the charge leaves the condenser, the voltage, which is proportional to the charge, begins to drop. The rate of discharge of the condenser, then, is proportional to its charge and both progressively decline. In quantitative terms the voltage V is equal to Q/C, where Q is the charge and C, the capacitance. Since $V = Q/C$ is the voltage drop IR across the resistance R,* while dQ/dt denotes the current I in the resistor, we have $IR = R\,dQ/dt = -Q/C$ and:

$$dQ/dt + Q/RC = 0 \qquad (2)$$

The negative sign occurs because IR and Q/C are in opposition. We divide the equation through by C and make the substitution $V = Q/C$, obtaining:

$$dV/dt = -V/RC \qquad (3)$$

The relation in terms of V is preferred because V is easily measured. If V is in volts and R in ohms, C will be in farads.

At this point we may compare eq. 3 with the equation for the washout of a simple one-compartment open system. This equation is obtained by inserting $S_2 \rightarrow \infty$ into the second of the two eqs. 1 in Chapter 1, which gives us $a_2 = 0$. Inserting zero for a_2 into the first of eqs. 1 in Chapter 1 yields

$$da_1/dt = -\rho a_1/S_1 \qquad (4)$$

If we compare eqs. 3 and 4, we note that there is a one-to-one correspondence between the voltage and the specific activity, between the capacitance and the compartment size, and between the conductance (reciprocal of the resistance) and the turnover rate. In general, this correspondence is the basis for the electrical analog method in tracer kinetics. Circuits are set up containing resistances and capacitances. Initial values of specific activity are simulated by discharging the condensers to zero and then applying specified voltages to some. The changes in voltage which then occur as a function of time are quantitatively determined, usually by oscillographic or other electronic recording techniques. These voltage changes are taken as analogous to specific-activity changes in a tracer system. This type of simulation is, of course, only one of many possible kinds. Inductances and currents can also be used, but capacitances and voltages have proved to be the most convenient.

Figure 23 shows illustrative electrical analogs for two types of catenary systems. It will be noted that simple exchange between two or more compartments can be simulated passively by connecting resistances

---

* In this chapter only, R will represent *electrical resistance* instead of tracer amount.

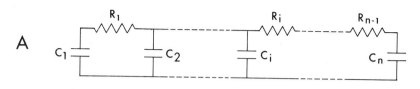

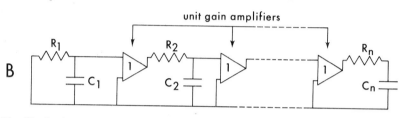

Fig. 23. Analogs for two types of catenary systems. A simulates pure exchange; B simulates one-way turnover.

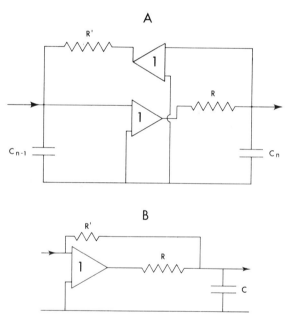

`Fig. 24.` Simulation of transport at non-equal rates in opposing directions, A by opposing units, and B with a single amplifier. It is assumed for simplicity that the source providing the input does not interact with the system. In A, $1/R$ and $1/R'$ are analogous to the individual opposing transport rates. In B, C is analogous to S_2 (S_1 not indicated), $1/R$ is analogous to the rate into compartment 2, $1/R'$ is analogous to the exchange rate between 1 and 2.

between two or more condensers. Thus, in the analysis of many closed steady-state systems, electronic amplifiers may not be required. (Of course, we except from this statement the amplifiers contained within oscillographic or other electronic display equipment.)

When we proceed to systems in which one-way transport such as turnover or washout occurs, then it is possible for an upstream compartment to influence a downstream compartment without any counter influence occurring. Here the device used must be one which, when an input voltage e is applied, will deliver an output voltage e, yet will not permit the output to affect the input. Such a function can be performed by a unit-gain amplifier, which is inserted between the corresponding resistance-capacitance stages (see Fig. 23B). It is, of course, possible to simulate any single transport rate between compartments in this way.

Therefore, in the case of pure exchange two amplifiers can be connected in parallel with opposing gains (Fig. 24A). Usually, when there are two non-equal opposing transport processes, a by-passing resistance can be used together with one amplifier to take into account the amount by which one rate exceeds the other (Fig. 24B). In the case of turnover plus exchange the conductance of the bypass ($1/R'$ in Fig. 24B) will be analogous to the exchange component.

Verification of the Analog Equations

Consider the network in Fig. 25A, and compare the voltage V_i of condenser i with that of condensers $i-1$ and $i+1$ (the latter not shown). We have

$$V_{i-1} - V_i = I_{i-1}R_{i-1}$$

where I_{i-1} is the current in the resistance R_{i-1} joining condensers $i-1$ and i. The rate of change of the charge on condenser i is the difference between the current flowing in, taken as positive to the right, and the current flowing out, i.e.,

$$dQ_i/dt = I_{i-1} - I_i$$

In terms of voltage rate of change

$$dQ_i/dt = C_i\, dV_i/dt$$

Combining the above relations, and noting that

$$V_i - V_{i+1} = I_iR_i$$

we obtain
$$C_i \frac{dV_i}{dt} = \frac{V_{i-1} - V_i}{R_{i-1}} - \frac{V_i - V_{i+1}}{R_i} \qquad (5)$$

If we compare this equation with eq. 1 in Chapter 5, where $\epsilon = 0$, we note the correspondence between V_i and a_i, between C_i and S_i, and between ρ_i and $1/R_i$. Thus the network in Fig. 25A is analogous to a catenary system under conditions of simple exchange.

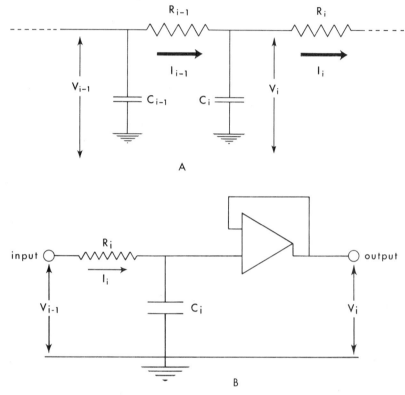

Fig. 25. Derivation of equations for catenary analogs. A applies to simple exchange; B applies to one-way turnover. In B, the triangle represents a unit-gain amplifier.

The analogy where $\rho_i = 0$ and $\epsilon \neq 0$ can be shown to be the circuit in Fig. 25 B. The current can flow only through the resistor into C_i, since the amplifier carries no current but merely propagates its input voltage into its output. The potential difference across R_i is

$$V_{i-1} - V_i = I_i R_i = R_i \frac{dQ_i}{dt} = R_i C_i \frac{dV_i}{dt}$$

thus

$$\frac{dV_i}{dt} = \frac{V_{i-1} - V_i}{R_i C_i} \tag{6}$$

A similar analogy is found in this circuit between V_i and a_i, between C_i and S_i, and between ϵ and $1/R_i$. Here, however, in the case of turnover in a catenary system, we note that, if S_i is to remain constant, the ϵ is the same between all compartments. For this reason it is more proper perhaps to consider the analogy between the "time constant" $R_i C_i$ of a given stage and the "time constant" S_i/ϵ of the tracer system. Thus the time constant $R_i C_i$ is varied by changing either R_i or C_i or both. Further information on this important subject may be found in the literature.

We may again note at this point an important fact relating to electronic analogs. As noted in Fig. 24, it is possible to connect unit-gain amplifiers in opposition in order to simulate exchange. By using different resistance values one may propose to simulate situations where transport rates are opposite but not equal. However, if unequal transport rates occur, compartmental contents will not necessarily remain constant. Thus, the steady-state situation no longer holds.

From time to time, workers have claimed to electronically simulate tracer experiments with unequal rates. However, in some of these situations, we suspect that the combination of rates so selected is such that compartment sizes should be changing with time. Such a situation cannot be simulated with electronic circuits involving fixed condensers whose capacitances cannot be varied. It is possible, of course, to simulate steady states if one of the opposing rates always exceeds the other by the same amount (exchange plus turnover).

Quantitation in Simulation: The Analog Computer

Simulation of tracer experiments by analogs is of interest qualitatively, but what is really desired is a quantitative expression of specific activity changes. Therefore, if this expression can not be obtained from the analog, it will still be necessary to solve the differential equations. However, by appropriate choice of the units of measurement, voltages may be converted to specific activities and other simulation parameters related to system parameters under study. In this way, the system becomes an "analog computer."

Of course analog computers may be more or less precise with a corresponding variation in price. As with a slide rule (also an analog computer), it is often sufficient to use a computer of moderate to low precision. In some instances, we sacrifice precision to obtain high speed. In this way solutions may be rapidly generated and the effect produced on them by varying input parameters more readily investigated.

Operational units and essential components of electronic analog computers. For a comprehensive discussion of analog computers, reference must be made to special books on the subject (52). However, for tracer kinetic problems in compartmental systems, the

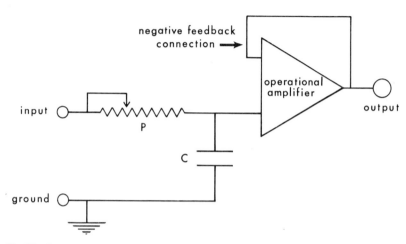

Fig. 26. Connections for a typical "lag" unit showing unit-gain amplifier, potentiometer P and condenser C.

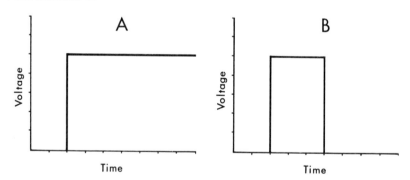

Fig. 27. Step, A, and square wave, B, voltages used in analog simulation.

situation becomes quite simple. The basic element of such a system is a combination of a resistance-capacitance unit and a unit-gain amplifier (Fig. 26). There are alternative ways in which such elements can be connected, but the net result is always the same. If a step voltage (Fig. 27) is applied to the input of such a unit, the output voltage will rise to the same ultimate plateau. However, the rise will be more gradual, in accordance with the following law:

$$E = E_\infty(1 - e^{-t/RC}) \tag{7}$$

Such a unit has received various designations: since it responds to a step voltage but in a sluggish or lagging fashion, it is referred to in some publications as a "delay unit" or "lag unit." The term lag unit will be adopted here, and the designation L will be used in block diagrams. This unit is one of the typical "operational" units used in analog computers.

Although it will be useful to understand the manner in which the lag unit functions internally, the main practical fact for the tracer worker is that a step applied to the input yields a lagging step at the output. By varying the time constant (by changing R or C or both), we can vary the rate of rise at the output; in fact, practical lag units are usually provided with such an adjustment. When, after the step has been applied for some time, the voltage is suddenly changed to zero, the output also falls but, again, in an exponential fashion. As the period between application and removal of input voltage shortens, we observe the typical response to a square-wave input (Fig. 28A). Finally, as the duration of the square wave becomes very short, the response following an initial sharp rise becomes proportional to $e^{-t/RC}$ (Fig. 28B), which is the response to a "delta function" (see Chapter 9, pages 172–175).

The latter response is an idealization which is never actually reached in practice. We may define a delta function as an impulse of infinite height and infinitesimal duration. In practical terms, however, if the duration of the impulse is quite short when compared to the response time of the lag unit, the result will conform quite closely to the idealized result. Such principles, are, of course, important in simulating the initial labeling of a system by delivering a short burst of highly active tracer into one of the compartments.

Given an input of labeled material, if we now compare the response of a single lag unit with the response of a one-compartment system such as the total body water, it is seen that the two are analogous. Since electronic pulse generators of versatile properties are available, the advantage of the analogy becomes apparent. By applying varying kinds of input pulses with different wave forms, we can readily simulate the response of the body water to arbitrary variations in the level of ingested labeled water. These variations might in one case be the result of chance environmental variations, for example, those in a study of tritium toxicity. In another case they may be the result of deliberately imposed input conditions in a controlled experiment. Clearly then, in addition to output-display equipment and one or more lag units, a device for providing appropriate input signals must be included in an analog computer. For simple step or impulse function input, conventional pulse generators are available. However, for most unusual inputs some type of arbitrary function generator may be required. A number of such devices have been described (52)

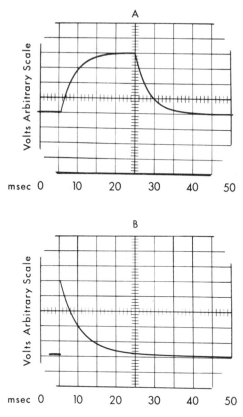

Fig. 28. Recorded response of a lag unit, A, to square wave and, B, to "delta function," or "spike," stimuli.   Time constant in A is 3.6 msec and in B 5.8 msec.   (Photographed from the screen of a Dumont Model 403 oscilloscope.)

including function generators which deliver an output voltage y which is a function of some input voltage x. Although function generators are convenient for some types of mathematical problems, most tracer simulations will be adequately treated by less-expensive units which deliver y as a function only of time.

Initial conditions.   In a number of instances, at the end of each cycle the input function may be chosen so that the output voltage of the lag unit is zero.   Other times, it is necessary to actually discharge the condenser of the lag unit to zero.   Since this discharge will not usually be instantaneous, analog computers commonly require a short conditioning interval before the actual simulation proceeds.   During this interval electrical "clamping" circuits cause the condensers of various operational units to be discharged to zero voltage. The simplest type of clamper is merely a relay whose contacts are connected across the capacitance. Other more elegant clamping methods have been described (52). Condensers which must be discharged are those which correspond to compartments having zero initial specific activity. Other condensers having a prescribed initial specific activity must be charged to the appropriate initial voltage. This charging may be done by the application of appropriate step voltages.

Other operational units. In the simulation of tracer kinetics there are various other operational units which will usually be employed. When the sum of two or more input voltages is applied to a unit, an adder must be used (Fig. 29, top). In block diagrams this unit is normally designated by A. Since particular circuits may have the property of inverting the signal, an inverter I must be employed if the signal must be reinverted (see Fig. 29, center). This unit-gain device does not change the scale of the signal. Finally, if multiplication of an input voltage by a constant is required, this operation is performed by a "coefficient unit" C (Fig. 29, bottom).*

Because of the limitations of electronic design, operational units possess a limiting maximum input and output voltage which cannot be exceeded if they are to function correctly. In such cases it is usually possible to keep the variable "in range" if we scale the voltages down by the use of C units in the input and then restore the original scale at a later point by similar means. The most convenient operational units available commercially are those which contain a warning device which is actuated when a voltage goes out of range.

Another type of unit which may sometimes appear in an analog computer is the integrator unit. Integrators are usually designated by J in block diagrams. They will deliver at their output the integral of the input signal. If the input signal be $V_i(t)$, the output signal will be

$$V_o(t) = K \int_a^t V_i(t) \, dt, \tag{8}$$

where K is a scale factor which is usually adjustable for the problem at hand. The lower limit a is usually zero although circuits are available permitting non-zero a values. Although the integrator is not necessary for simulation of tracer kinetic problems, occasions arise when its use is convenient. For example, J units are of use in simulating experiments

* Here and in following sections of the present chapter, we will have occasion to refer to "C units" and to C_1, C_2, etc. This notation is chosen because it is frequently used in the analog computation literature. The distinction between C for "coefficient" and C_i, used elsewhere for "capacitance i," should be kept clearly in mind.

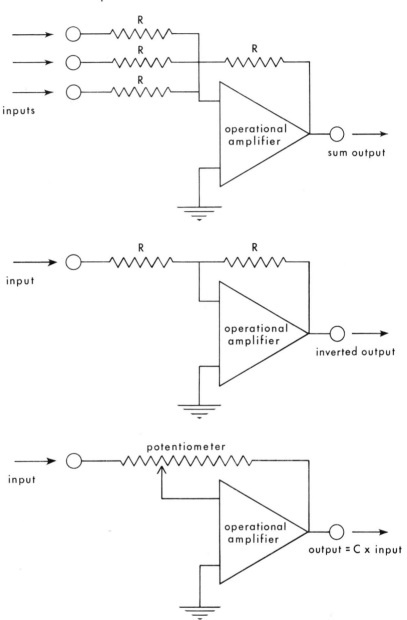

Fig. 29. Typical operational units employed in analog simulation. In the adder (top) the output voltage is the negative sum of the inputs. Negative signals are changed to positive in the inverter (middle). Scaling (multiplication by a constant with inversion) is performed by the coefficient unit (bottom).

such as kinetic studies in the mammalian circulation. Here, the amount of label in an organ may be represented by the integral over time of an arterio-venous difference. More will be said on this subject in Chapter 10.

Practical Considerations in Analog Computation

Scale factors and time constants. Whatever the system used, we must establish a one-to-one correspondence between the units of measurement as well as between the variables being measured and simulated. Both the tracer system and the analogs normally used involve the independent variable of time. Whether the analog is hydrodynamical, electronic, or of some other type, the time scale of the tracer system could be the same as the analog system, but usually it will be different. The first step, then, is to decide what the relation must be between the time scales of the two systems.

Consider a one-compartment system such as the one involved in the study of body water turnover. Here it may take perhaps 10 days for half of the initially labeled water to be replaced by unlabeled water. Equation 4, in Chapter 1, expresses the system fairly accurately. The exponential constant ρ/S for the system yields the reciprocal of the time for the specific activity to be reduced by a factor $1/e$, where e is the base of natural logarithms. The time for reduction by $1/2$ or "half-value time" is more commonly used and is $0.693S/\rho$.

Thus, for body water, S/ρ is about 14.4 days, computed from the half-value time which is the experimentally determined "biological half-life." In simulating this system, we may prefer to deliver a voltage "spike" to a lag unit with adjustable resistor. By observing the response, we may then be able to set R and thus the product RC so that days in actual time coincide with convenient scale divisions of computer time.

One advantage of the electronic analog computer is the ease with which it will yield a direct plot of voltage versus time through the use of display equipment, such as an oscilloscope, a recording galvanometer, or a servo-driven X-Y plotter. In obtaining such a plot, we can use the expected duration of the experiment to determine what portion of the curve will be displayed. However, the flexibility of electronic devices permits ready alteration of time scales as desired. In the choice of scale, we may suppose, for illustrative purposes, that a computer of the high-speed repetitive type is being used. The results are being continuously displayed on the screen of a cathode-ray oscilloscope at the rate of 20 cps.

A study of body-water kinetics, in which the curve of tracer specific activity is to be followed for two half-value times, would require each analog display to proceed for two analog half-value times. Since in the present case one sweep, representing 1/20 sec or 50 msecs, must include two half-value times, then each half-value time would be 25 msecs. Thus the RC value would be $0.025/0.693 = 0.036$ sec. A resistor of 0.1 megohms and a capacitance of 0.36 microfarads would then be a correct combination. If the horizontal scale of the oscilloscope were graduated in tenths, then each division would represent 1/5 half-value time and be analogous to 2 days of the tracer experiment. Of course if each sweep requires a 10-msec conditioning period for clamping, the scale will be correspondingly changed or the simulation will occur over a slightly shorter interval.

Connection of operational units. The manner in which operational units may be connected to simulate a kinetical problem may be illustrated by a catenary system with uniform turnover. Consider the lag unit which might be the ith member of a catenary system (Fig. 23B). The unit itself, if isolated, will follow the equation

$$C_i dV_i/dt = -V_i/R_i \tag{9}$$

If there is no input signal from the preceding stage, the voltage at the output of the unit is

$$V_i = -R_i C_i dV_i/dt$$

However, the presence of an input signal $V_{i-1}(t)$ causes the current in the resistor to be proportional to the voltage difference across the resistor. Thus

$$V_{i-1} - V_i = R_i C_i dV_i/dt$$

We must therefore consider the output voltage as dependent upon both the relaxation properties of the unit and the input signal. From this point of view the behavior of a compartment in a tracer kinetic system can be considered as the response of the compartment to the effects of all compartments which can deliver label to it.

Consider, then, the general case of the ith compartment in a multi-compartment system. Rewriting eq. 2 in Chapter 3, we have

$$\frac{da_i}{dt} + a_i \sum_j \frac{\rho_{ji}}{S_i} = \sum_j \frac{\rho_{ij} a_j}{S_i} \tag{10}$$

If the right hand side is zero, then the compartment will relax according to the equation

$$\frac{da_i}{dt} = -a_i \sum_j \frac{\rho_{ji}}{S_i} = -K_i a_i \tag{11}$$

whose solution is $a_i = a_i(0)e^{-K_i t}$ where

$$K_i = \sum \rho_{ji}/S_i$$

We may therefore simulate compartment i by a single lag unit whose RC value corresponds to the K obtained by taking the sum of all outgoing transport rates and dividing by the compartmental contents.

In this way, we may select lag units for all compartments and, by appropriate choice of scale factors, adjust their RC values for the system under consideration. Having matched the units in terms of their zero input response, we must then consider their interactions. Of course the input to each unit must take into account label entering a compartment from all compartments which can donate to it. Thus the input to a unit must include the sum of outputs from all corresponding lag units j represented by the right side of eq. 10. This requirement is met by connecting the outputs of these units j to the input of unit i through an adder. However, the various units will have different amounts of effect. Some will donate labeled material at a slow rate because their transport rates into compartment i will be slow. The variable input conditions are taken into account by including in the output of each unit a coefficient unit C.

The factor to be set on each C unit may be determined by referring to eq. 10. As $t \to \infty$, the derivative becomes negligible. At this point the specific activity is equal to the weighted mean of the specific activities of all donating compartments. The effect of each in compartment i is reflected in its weighting factor, which is the ratio

$$\frac{\rho_{ij}}{\sum_j \rho_{ji}} \tag{12}$$

This ratio is the factor which must be set on the C unit for the jth lag unit.

Summary of rules for connection of operational units. These principles may be summarized as follows:

A. Determine the relaxation time for each compartment in the system. This expression will be the total amount of the substance S in the compartment divided by the sum of all transport rates out of the compartment. If S is measured in milliequivalents and the transport rates in milliequivalents per second, the relaxation time will be in seconds. In general consistency among units must be maintained.

B. Establish a corresponding relaxation time for the "lag" units used to simulate the compartments. Usually this time, in seconds, will be equal to the product of the resistance of the unit in megohms and the capacitance in microfarads. These times may be chosen arbitrarily so that the time scale for the solution display is appropriate for the problem at hand.

C. Adjust the units for the correct RC value and test their response to a pulse or step. For problems in which the initial value in certain compartments is zero, "clamping" units must be applied to the condensers of the appropriate lag units so that they can be discharged to zero at the end of each display scan.

D. Make connections between analogous lag units for each compartment which can receive activity from other compartments. The output of the "donor" unit is connected to the input of the "recipient" through a coefficient unit. The percentage of unit gain to be set for the unit will be the ratio of that particular inward transport rate to the total transport rate out of the analogous compartment.

E. Apply connections through an adder unit for compartments which receive multiple inputs. For those computers in which addition occurs with a reversal of sign an inverter must be included in the outgoing connection of the adder.

F. In order to simulate the application of label to the system, apply signals to appropriate compartments. Simulation of constant input suddenly starting at zero time is achieved by applying a step voltage to the input of all lag units which receive constant specific-activity input (usually only one). For compartments which initially receive a burst of label, a pulse may be applied. If a sufficiently large and brief pulse is not available, a convenient alternative is to apply a brief step voltage during a short period before the solution is to be displayed while the clamps are applied to all units which are initially zero. This step voltage is then removed at the moment of release of the clamps. (The input unit should not disturb the impedance of the system to which it is connected.)

G. The specific activity of any compartment which may be of interest will be simulated by the voltage on the condenser of its appropriate lag unit. The voltage may be directly applied to any display unit such as an oscilloscope. (The display unit must not have an appreciable loading effect on the system.)

Engineering Considerations in Electric Analogs

There are a few practical considerations which may govern the investigator's choice of analog computation components. Of course, cost is one. Just as with another type of analog computer, the slide rule, the precision of the result is generally related to the number of dollars available and to the space limitations. Ordinarily, the ease with which differential equations for some, if not all, steady-state systems can be solved will tend to rule out the need for elaborate facilities used in analog solution. In this case the choice may well be the relatively inexpensive computer which has the flexibility, convenience, and ease of a pocket slide rule. As the number of compartments increases, the number of operational units required will increase. Therefore if one is to remain within a reasonable budget, less expensive components may be in order.

If analog computation is to be practically feasible, a zero input must yield zero output. This condition is normally obtained by some sort of balance adjustment in the amplifier of the unit. However, if lower precision and rapid computation is in prospect, an appreciable rate of "drift" of the zero setting may not prove objectionable. Nevertheless, selection of equipment must include a consideration of the specified drift rate. Frequently the more expensive amplifiers are especially stabilized against drift.

The gain of amplifiers requires careful control and a stable gain is generally obtained by the use of negative feedback. The inherent gain is made as high as possible, and a large amount of feedback is employed. In this way the feedback factor entirely controls the gain. This situation is preferable in the unit-gain amplifiers employed in lag units. Since the feedback factor is usually obtained by some type of voltage divider, it is good practice to use wire-wound resistances. In most cases such resistances are advantageously employed in all critical places in analog computers. Better still are the highly linear, multiturn potentiometers such as the Beckman Helipot. In the case of units with very long time constants large condensers become expensive. Although resistors with high resistance are available, they are not readily varied. Usually they may be connected to a tap switch for variation through large steps with a continuously variable lower resistance vernier connected in series.

Condensers used in analog computation should be of high quality. A poor condenser in a lag unit, for example, will cause the response to depart significantly from true exponential behavior. For this reason good quality capacitors with a "mylar," or polystyrene, dielectric are preferred. In recent years there has been a considerable reduction in the expense of such components.

An important consideration in most electronic devices is the input and output impedance. Operational units employed in analog computers will preferably have reasonably high input impedance and low output impedance. When they are designed in this manner, the output of one unit can be readily connected to the input of one or more other units with little risk of "loading" trouble. The small current necessary to maintain the input voltage at the correct applied value is readily delivered by the preceding stage without a loss due to internal generator resistance. Low impedance of all units to ground is favored if the circuits are to operate at high repetition frequency. Here a high impedance favors pickup of stray

60-cycle alternating-current voltages and also favors "cross talk" pickup between units.

Generally speaking, then, it is best to limit input impedances to less than 100,000 ohms to ground in these cases. Because of this fact, output impedances should be a few thousand ohms at most, and, if higher impedances are used, shielding may become necessary. Shielding can be successfully achieved against stray electrical fields but difficulties arise with stray magnetic fields. Low impedance circuits are particularly vulnerable here. Thus transformers and inductances containing alternating currents must be kept at a reasonable distance.

In choosing operational amplifiers some attention should be paid to the maximum value of input voltage which may be applied. The amplifier's ability to accept both positive and negative input voltage should also be considered because maximum versatility is achieved by units which will accept voltages of either sign up to 50 volts or more. In this way less difficulty in scaling the variables will be encountered. Units with maximum signal-to-noise ratio are preferred, since this ratio will determine the maximum possible precision.

Convenience in making connections and freedom from maintenance problems may be important considerations in the choice of analog computer equipment. These conditions will depend on the situation. One investigator may have a Ph.D. in electrical engineering and be willing to devote a considerable amount of time to adjusting and redesigning his equipment. To such an individual a set of amplifiers which can be operated with home-made modular plug-in units will have considerable merit. With these units an amplifier can be connected at will as an adder, as a lag unit, or as one of many other types of units. The expert will readily locate and correct loose connections and other pitfalls. Another investigator may prefer a computer with a well-designed problem board of permanent construction. Usually, for equal funds less versatility is obtained but more stability and simpler operation.

Some Typical Analog Setups

Since it is often possible to contrive simple electrical analogs from existing equipment, a brief description of two typical non-commercial research setups will be given. Both were built in this writer's laboratory at different times.

The first analog was used for simulating electron scattering and heat flow. Also tracer-kinetic problems were studied in simple steady-state systems of the pure exchange type. A vibrating reed electrometer, often found in a laboratory engaged in research with C^{14}, was available (Applied Physics Corporation Model 30). The electrometer was already connected to drive a recording potentiometer. A series of well insulated polystyrene condensers was mounted in a small sheet-metal box. Connections among them were made by vacuum sealed resistors (Victoreen Instrument Company, Cleveland, Ohio). Initial charge was applied as desired through a battery and key.

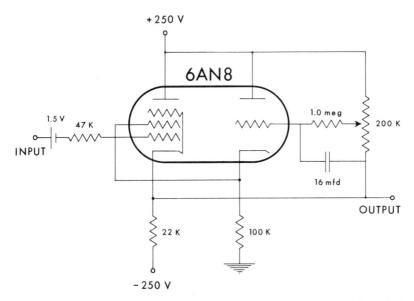

Fig. 30. Unit-gain amplifier circuit modified from Brownell, Cavicchi, and Perry (51).

The vibrating reed was connected through a mechanically stable, well insulated and shielded lead in order to record, in turn, the charge on various condensers as the system relaxed from the initial conditions. The time constants were of the order of several minutes and, with condensers of 1/4 microfarad, resistances of 20 or more megohms were required. For this reason care was taken to avoid leakage to ground and stray electrostatic disturbance. With a system of this sort successful simulation of exchange of K^{42} between erythrocytes and plasma was accomplished. Other tracer experiments were also simulated.

A second type of analog was developed to simulate circulatory mixing and also washout of a catenary system. A series of ten operational amplifiers was constructed (Fig. 30) at a cost of about $10.00 per amplifier. They were supplied by a conventional well-stabilized power supply and connected as a series of lag units with time constants of approximately

0.05 sec each (Fig. 31). The condenser of the first unit is charged by connecting a small priming condenser to it. This small condenser C_1 is first applied to a battery. After the ground connections which "clamp" the system to zero initial voltage are removed, the switch is thrown and C_1 shares its charge with the condenser of the last unit which is the input to the first. The input of a Sanborn Physiological Recorder (Twin Viso) with a D.C. preamplifier was connected to various condensers to record the transient response of the system. It is hoped that the descriptions of these two readily devised units will help remove any air of mystery which may surround the adaptation of existing equipment to simulation of tracer kinetic problems.

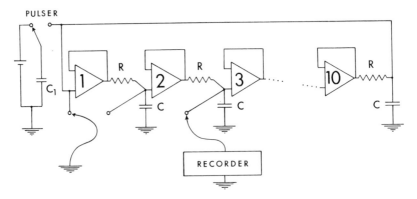

Fig. 31. Ring connection of operational units for simulation of circulatory mixing and of washout in catenary systems. Grounding the condensers simulates initial removal of tracer from the system. On removing the ground connection, tracer injection is simulated by throwing the switch to connect C_1 to the system as shown.

Those who plan the serious use of analogs will perhaps derive benefit from a description of a more versatile arrangement built mainly around commercial components.* Details of this and other types of commercially available equipment may be found in the literature (52). The computer is shown in Fig. 32. The principal feature consists of three Model HK Operational Manifolds (upper right) supplied by a Model SR 400 Power Supply. Each of these manifolds contains ten operational amplifiers which, by the use of Model K modular plug-in units, may be arranged to perform various operational tasks. The units are driven by a Philbrick Model CS-2 central-signal component which emits square pulses of variable voltage and duration; sawtooth signals of variable slope; and constant

* Obtained from George A. Philbrick Researches, Boston, Mass.

Fig. 32. A typical medium-price analog computer installation combining home-made components with Philbrick units. Right-hand rack contains (top to bottom): three manifolds of ten operational amplifiers each, one "central-signal" component, a power amplifier to operate clamping relays, and a power supply. Modular plug-in units project from the front of the operational manifolds. The left-hand rack contains a set of mounts for making passive connections to plug-in modules, a function generator, a two-channel multiplier-divider unit, a Philbrick "central-response" unit and a unit for producing high-voltage pulses (see Fig. 34). To the left of the assembly is a cathode-ray oscilloscope used to display the computer output signal.

d-c voltages. Repetition frequency is variable from 20 cps to less than 0.2 cps. Signals are emitted for the control of clamping devices, and, in addition, a signal is generated for driving the horizontal sweep of a cathode-ray oscilloscope, which is normally used for displaying the solutions.

There are several additional features which are convenient, although not essential, for basic kinetic studies in steady-state compartmental systems. A Philbrick Model CRM central-response unit permits five input signals, taken from various locations in the simulated system, to be simultaneously displayed on the oscilloscope. The fifth signal may be taken at will from 4 additional locations. The resulting output display is superimposed on an ingenious electronically generated grid which appears on the oscilloscope face. It should be noted that this device requires an oscilloscope whose axis modulation is such that a positive signal intensifies the trace.

Other added features are a pair of two channel Model MU/DV multiplier-divider units by which quotients or products of two variables can be obtained. Finally, an excellent feature, although costly, is a Model FFR function-generating component. This device produces an output voltage which, in a rather complex manner, can be related to the input under arbitrarily adjustable control from a series of knobs on the front. Usually an experimenter will use, for the input voltage, a sawtooth relation which will vary linearly with time. If this is connected to the horizontal sweep of an oscilloscope as well, the output can be displayed as a vertical deflection. By appropriate adjustment a wide range of curves can be obtained corresponding to a wide choice of output-wave forms.

A typical use of the function-generating component is to connect one or more operational units to simulate the activity in the thyroid of a sheep which might be exposed to some arbitrary time distribution of radioactive fallout containing I^{131}. The thyroid may be approximated for this purpose by a single lag unit. After observing the effect of single pulsatile exposures and adjusting the "biological half life" to 12 days, we are able to predict the response to continuously variable fallout. Voltage curves are synthesized with the FFR unit representing varying levels from day to day. This variable input function is fed into the single lag unit replacing the test pulses. Figure 33 shows a typical result of such a simulation, where **A** is the fallout "stimulus" and B the system "response."

It is occasionally advantageous to have a source of sharp high-voltage pulses for analog computer work. An experimental circuit developed for this purpose is shown in Fig. 34. Pulses of 800 to 900 volts and approximately 20 microseconds duration are emitted. These can be synchronized by the clamping impulse of a Philbrick CS 2 unit. Internal adjustments (P_1 and P_2) permit the interval between the removal of the clamps and the

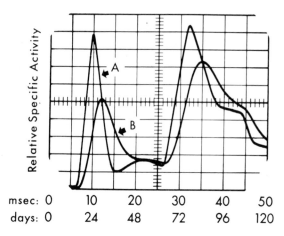

Fig. 33. Simulation of response of a biological system to variable input. Curve A is a hypothetical variable level of radioactive iodine (fallout). Curve B is response of "sheep thyroid gland." Horizontal coordinates are given both in true time and computer time scales.

delivery of the pulse to be varied to prevent their overlap. Figure 28B shows a typical response of a single lag unit to the impulse from this device. In the construction of the device, stability was obtained by careful isolation of amplifiers V_1 from the high-voltage components.

Analog Solution of a Typical Problem

In the field of iodine metabolism tracer methods have been of particular importance, and as a result, analyses based on kinetics of compartmental systems have proved of value. Analog methods have been of some assistance in the kinetic analysis. We have selected as an example a model by Stanbury et al. (53) of iodine metabolism studied in a human subject. This example was investigated by Brownell, who used analog methods. The assumed compartmental model with assumed rates and contents is shown in Fig. 35. We refer to eq. 2, Chapter 3, and designate the infinite environmental compartment as 0 with constant zero specific activity.

If we insert $a_i S_i$ for R_i and factor S_i out of the derivative, the differential equations for this system are:

$$32 da_1/dt = -(54 + 15 + 55)a_1 + 70a_4$$
$$300 da_2/dt = 15(a_1 - a_2)$$
$$50 da_3/dt = 55(a_1 - a_3)$$
$$1200 da_4/dt = 15a_2 + 55a_3 - 70a_4 \qquad (13)$$

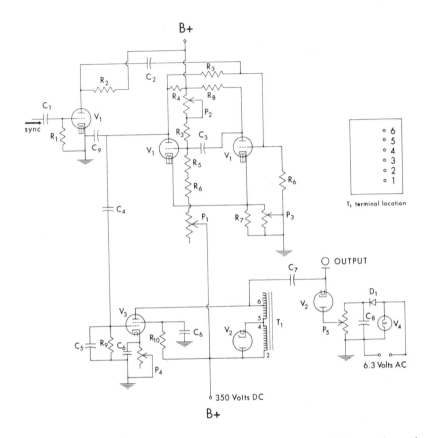

Fig. 34. Circuit diagram for a high-voltage pulser. Increase V_2 cathode-ground capacitances with short, twisted lengths of well-insulated wire as trimmers. Experiment with optimal placing of leads and grounds.

$C_1 = 600\ \mu\mu\text{f}$	$R_1 = 100$ kilo ohms	$V_1 = 1/2$ 12AU7
$C_2 = 270\ \mu\mu\text{f}$	$R_2 = 5.6$ kilo ohms	$V_2 = 6V3$
$C_3 = 0.0066\ \mu\text{f}$	$R_3 = 1$ megohm	$V_3 = 6AV5$
$C_4 = 0.1\ \mu\text{f}$	$R_4 = 18$ kilo ohms	$V_4 = \#51$ pilot light
$C_5 = 0.002\ \mu\text{f}$	$R_5 = 130$ kilo ohms	$P_1, P_2 = 0.5$ megohm
$C_6 = 20\ \mu\text{f}$	$R_6 = 470$ kilo ohms	$P_3 = 0.25$ megohm
$C_7 = 175\ \mu\mu\text{f}$ H.V.	$R_7 = 10$ kilo ohms	$P_4 = 1000$ ohms
$C_8 = 500\ \mu\text{f}$	$R_8 = 180$ kilo ohms	$P_5 = 1.0$ megohm
$C_9 = 25\ \mu\mu\text{f}$	$R_9 = 0.47$ megohm	$D_1 = $ RE 40 diode
	$R_{10} = 56$ kilo ohms	$T_1 = $ Thordarson fly 121

Compartment 1 is initially labeled. Let its initial specific activity be $a(0)$. To obtain the specific activity relations for each compartment we take Laplace transforms,

$$(32s + 124)\alpha_1 - 70\alpha_4 = 32a(0)$$
$$-15\alpha_1 + (300s + 15)\alpha_2 = 0$$
$$-55\alpha_1 + (50s + 55)\alpha_3 = 0$$
$$-15\alpha_2 - 55\alpha_3 + (1200s + 70)\alpha_4 = 0 \qquad (14)$$

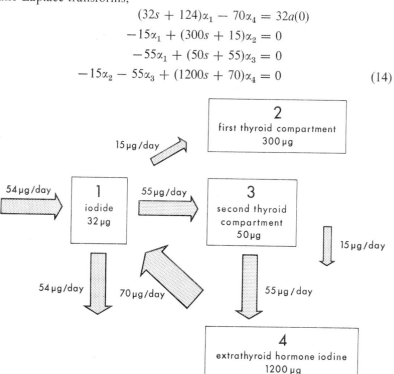

Fig. 35. A typical tracer compartmental system. This model was one of a series postulated by Stanbury et al. to represent iodine metabolism in human subjects observed in an endemic goiter area. The original electronic simulation was performed by Brownell. Redrawn by permission of Harvard University Press and the World Health Organization.

The denominator determinant Δ has the form

$$\begin{vmatrix} 32s + 124 & 0 & 0 & -70 \\ -15 & 300s + 15 & 0 & 0 \\ -55 & 0 & 50s + 55 & 0 \\ 0 & -15 & -55 & 1200s + 70 \end{vmatrix} \qquad (15)$$

and equals $5.76 \times 10^8 s^4 + 2.928 \times 10^9 s^3 + 2.76732 \times 10^9 s^2 + 2.100255 \times 10^8 s + 3.1185 \times 10^6$.

The roots of Δ can be shown by substitution to be (to sufficient approximation)

$$\lambda_1 = -0.02002 \qquad \lambda_3 = -1.137$$
$$\lambda_2 = -0.06155 \qquad \lambda_4 = -3.864$$

The transforms of the solutions are expressed in the following way:

$$\Delta\alpha_1 = 32a(0)[(300s + 15)(50s + 55)(1200s + 70)]$$
$$\Delta\alpha_2 = 32a(0)[15(50s + 55)(1200s + 70)]$$
$$\Delta\alpha_3 = 32a(0)[55(300s + 15)(1200s + 70)]$$
$$\Delta\alpha_4 = 32a(0)[15^2(50s + 55) + 55^2(300s + 15)] \qquad (16)$$

where Δ is the denominator determinant.

By partial fractions we then have the equations:

$$\frac{\alpha_1}{a(0)} = \frac{0.00695}{s + 0.02002} - \frac{0.000227}{s + 0.06155} - \frac{0.013395}{s + 1.137} + \frac{1.0068}{s + 3.864}$$

$$\frac{\alpha_2}{a(0)} = \frac{0.0116}{s + 0.02002} + \frac{0.000983}{s + 0.06155} + \frac{0.000616}{s + 1.137} - \frac{0.01320}{s + 3.864}$$

$$\frac{\alpha_3}{a(0)} = \frac{0.00708}{s + 0.02002} - \frac{0.00024}{s + 0.06155} + \frac{0.39401}{s + 1.137} - \frac{0.40064}{s + 3.864}$$

$$\frac{\alpha_4}{a(0)} = \frac{0.01226}{s + 0.02002} - \frac{0.000393}{s + 0.06155} - \frac{0.016743}{s + 1.137} + \frac{0.004868}{s + 3.864}$$

Inversion (Fig. 36) yields the following results:

$$\frac{a_1(t)}{a(0)} = 0.00695e^{-0.02002t} - 0.000227e^{-0.06155t}$$
$$- 0.013395e^{-1.137t} + 1.0068e^{-3.864t}$$

$$\frac{a_2(t)}{a(0)} = 0.0116e^{-0.02002t} + 0.000983e^{-0.06155t}$$
$$+ 0.000616e^{-1.137t} - 0.0132e^{-3.864t}$$

$$\frac{a_3(t)}{a(0)} = 0.00708e^{-0.02002t} - 0.00024e^{-0.06155t}$$
$$+ 0.39401e^{-1.137t} - 0.40064e^{-3.864t}$$

$$\frac{a_4(t)}{a(0)} = 0.01226e^{-0.02002t} - 0.000393e^{-0.06155t}$$
$$- 0.016743e^{-1.137t} + 0.004868e^{-3.864t}$$

Obviously much labor is involved in obtaining such a result. First it is necessary to solve a quartic equation to obtain the λ's. Then a considerable amount of arithmetic and algebra intervenes before the transforms are obtained in their final form. It is in problems of this sort that electronic analogs show their value. The appropriate connection scheme is indicated in Fig. 37. The time scale is chosen so that 20 days corresponds to 40 msecs, which in the case of a Philbrick computer is the time for one complete oscilloscope sweep after release of the clamps. Table 2 yields the time constants and circuit parameters for the lag units.

The choice of resistance or capacitance is arbitrary, since only the product must be specified. Practically, low resistors reduce the problem of a-c pickup. However, the larger condensers are expensive.

Another problem may also be encountered with large condensers in cases such as this one. After each sweep, the system must be returned to

Table 2
Time constants and circuit parameters

Compartment	Total outgoing rates, μg per day	Time constant, days	Resistance, megohms	Capacitance, μf	$R \times C$ msec
1	124	0.258	0.0469	0.011	0.516
2	15	20	0.40	0.10	40
3	55	0.91	0.1085	0.0167	1.82
4	70	17.1	0.0855	0.4	34.2

zero initial conditions in all compartments except the first. This requirement necessitates connecting clamping circuits to compartments 2, 3, and 4. Of course it may be sufficient to clamp only the compartments of large time constant. If large condensers are to be efficiently discharged, the device used for discharging the condensers must carry a greater load. If this device is a high-speed relay with small contacts a limit may be set on the condenser size.

The final step is to determine the settings for the C units. Coefficients C_2 and C_3 will provide a general illustration. It is necessary to simulate 15 μg per day entering compartment 4 from compartment 2. This amount represents 15/70 of the total amount leaving compartment 4, so unit C_2 is adjusted to reduce the output voltage by 15/70 or 21.4 per cent. Similarly C_3 and C_4 are set at 78.6 per cent and 56.3 per cent respectively. In compartment 1, material of specific activity zero may enter and leave the compartment. This could be simulated by a C unit set to zero but that

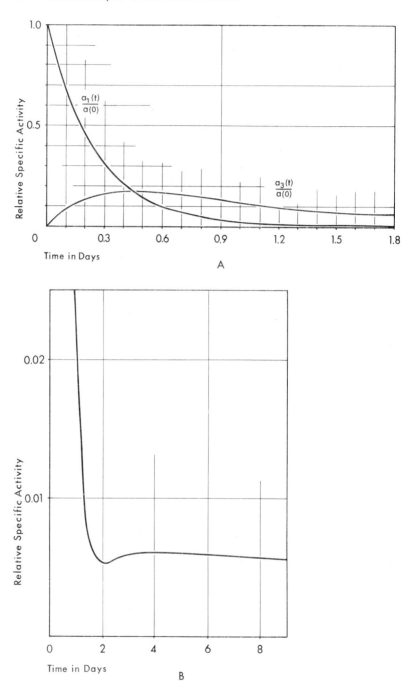

A

B

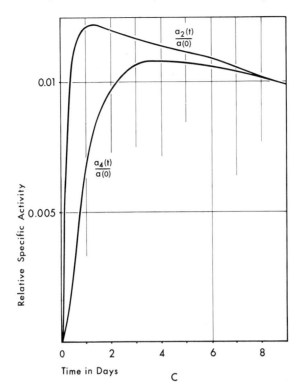

Fig. 36. Solutions for the system of Fig. 35. A: Compartments 1 and 3; B: lower portion of $a_1(t)/a(0)$ on an expanded scale; C: compartments 2 and 4 on an expanded scale.

would be equivalent to no connection at all. In such cases, the sum of the settings of the C units connected to a compartment input will be less than 100 per cent. The difference represents non-radioactive material which will add nothing to the amount of tracer entering the compartment but merely act to dilute and wash tracer out. In compartment 1, 54 μg per day of unlabeled iodide is entering compartment 1. This represents 43.7 per cent of the total flux through the compartment. No electrical representation will be required in the analog connections.

It is sometimes necessary to obtain a presentation of the total activity of two or more combined compartments. In the case of I^{131} the available methods of measurement do not provide separate measurements of all the subcompartments which might exist in the thyroid. Two compartments were postulated by Stanbury et al., to exist in the patient which the model in Fig. 37 was taken to represent. Here it is possible, as shown in the figure, to connect the outputs of compartments 2 and 3 through C and C′

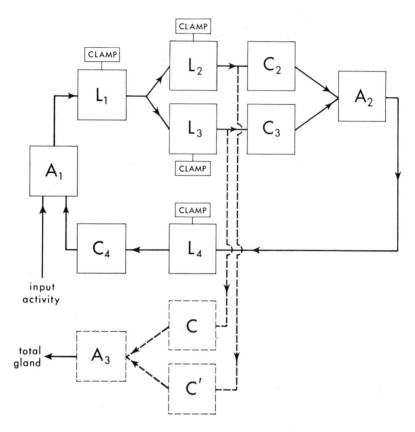

Fig. 37. Electrical analog connections for simulation of compartmental system of Fig. 35. L_1 to L_4 are lag units representing compartments 1 to 4. Connections are made through adders A, and coefficient units C. Weighted mean values for compartments 2 and 3 are obtained by A_3, C and C′ (dashed connections).

units and A_3 in order to yield an arbitrary weighted mean for oscilloscope display.

Theoretically in the case of compartment 1 the simulation could be achieved by a standard lag unit. Actually the pulser in Fig. 34 reflects some capacitive reactance into the unit. For this reason a special unit (Fig. 38) was used. It was also convenient to combine A_1 and L_1 in the same unit. The time constant was arbitrarily adjusted by varying P_1.

To obtain appropriate initial conditions in compartment 1 several procedures may be used. The compartment may be clamped and then a sharp pulse applied at the moment of unclamping which is understood to be "zero" sweep time. An alternative is to apply a conditioning voltage through A_1 during the clamping of the other units. This voltage is then removed at zero time. Since the pulsing unit of the Philbrick computer yields zero voltage during clamping and also delivers the leading edge of the pulse at zero time, it is necessary to apply constant direct current in conjunction with a negative pulse which just reduces the input to zero at the start of the solution display. The pulse may be broadened to include the entire 40 msec of solution time, but a somewhat narrower pulse may be necessary to enable L_1 to completely recover its initial condition before the clamping period is complete.

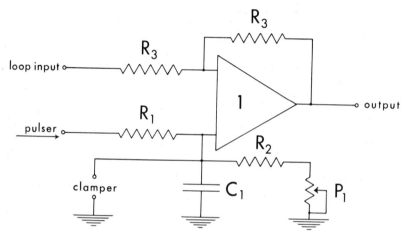

Fig. 38. Special lag unit for use with high-voltage pulse generator (see Fig. 34). Input from compartment 4 connected to "loop input." $R_1 = 0.22$ megohm, $R_2 = 56$ kilo ohms, $R_3 = 1.0$ megohm $P_1 = 0.1$ megohm, $C_1 = 0.004\ \mu f.$

Figure 39 shows a series of displays of representative results obtained from simulating the iodine compartment system. The voltage pulse applied to unit 1 established an initial level of 50 volts in the unit. The cathode ray traces indicate a few typical technical errors occasionally encountered in work of this sort particularly with the less expensive computers. Thus, for example, the slight waver in the horizontal component of Fig. 39A is due to extraneous disturbance caused in part by 60-cycle a-c hum. The high magnification was chosen to illustrate the details of the sharp dip in the curve at 2 to 3 msec, which is comparable to the dip in the computed curve, as in Fig. 36B. Although there is a real oscillation in the curve following this dip it cannot be readily separated from the interference. This is not to be considered as poor computer performance, however, as the interference component represents less than 100 mv. Since the initial voltage was 50 volts this is less than 0.2 per cent.

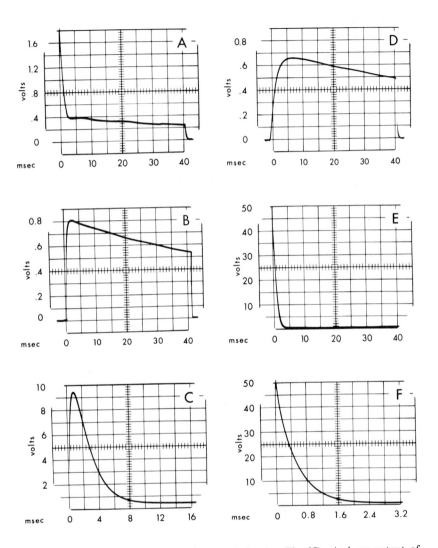

Fig. 39. Cathode-ray displays of iodine simulation (see Fig. 37). A shows output of unit 1 on magnified scale. In A, unit 1 is connected into the system. Unmagnified output of isolated unit 1 with two time scales is shown in E and F. B, C, and D show units 2, 3, and 4, respectively. Note, in comparing with Fig. 36, that the voltages must be divided by 50, the initial voltage on L_1. Also, 20 days in Fig. 36 corresponds to 40 msec here.

By applying the clamp at 41 msec, we were able to obtain a better determination of the zero setting for the display. This is crucial since the ordinate values of the curve are small at this time in unit 1 and precise determination is associated with some difficulty. The best value which could be obtained was 0.39 volts which corresponds to a relative specific activity of 0.0078. Reference to the computed value in Fig. 36B shows an error at approximately 0.0025. A probable origin for this error is a d-c offset in the unit of about 120 mv, which, although only about 0.25 per cent of the initial 50 volts, is a relatively serious discrepancy. The system being studied in this case consists of a central "plasma iodide" compartment which is only slightly affected by the other three components. Figure 39 (E and F) shows the gross behavior of the compartment when disconnected from the remainder of the system. When the complete system is simulated it is only through thirty-fold magnification, as in Fig. 39A, that evidence of departure from true exponential behavior is seen. It is clear that situations of this sort place a stringent requirement on computer precision in frequent instances.

Of course an error in the setting of the RC value of a lag unit can produce errors in computation. In Fig. 39B, as a result of such an error, the maximum in compartment 2 is 0.8/50 corresponding to $a_2/a(0) = 0.016$. This is approximately 30 per cent higher than the computed value in Fig. 36A. The general shape of the curve, however, is not seriously altered. The dominant effect of compartment 2 on 4 causes an increase in the compartment 4 maximum also. Another typical error is well illustrated in Fig. 39D, where the zero of the time axis is significantly offset by premature arrival of the voltage pulse in unit 1. A similar offset is suggested in Fig. 39A, where the dip theoretically should occur at 4 to 5 msec.

From these results it is clear that electrical analog methods require considerable care in their use. In setting the proper RC values for lag units it is essential to isolate and test them in situ with single pulses. The response curve of each isolated unit should yield a true exponential relation, and the half-value time should be precisely set. The system should be tested by simulating a few problems whose solutions have been determined analytically. At the same time analytical solution can be facilitated if certain features are indicated from analog simulation. The fact that compartment 1 is shown by the analog to behave as though it were practically isolated—Fig. 39 (E and F)—permits a ready approximation for compartments 2 and 3 since they are driven solely by 1. The methods for obtaining these approximate solutions as perturbations are discussed on pages 83–85 of the previous chapter. Generally speaking in kinetical computations analytical, numerical and analog methods tend to re-enforce one another.

1

Non-Steady-State Compartmental Systems

There are numerous systems which do not fit adequately into the steady-state model. In many of these, compartment sizes vary appreciably with time. In mathematically describing these systems we encounter difficulties, since solutions of systems of linear differential equations with non-constant coefficients cannot, in general, be obtained by standard procedures. Moreover, the fact that systems of this sort are in the form of differential-difference equations does not seem to provide any advantage in solution. The method of Laplace transforms is of little use. This fact may be appreciated if the method is applied to eqs. 12 in Chapter 4 for the non-steady state case.

If, in the steady state, the ρ_{ij} and the S_i on the right are constant, their quotient is also constant and may be factored out of the Laplace transform by rule B, page 50. The transformed equations then form a system of linear algebraic relations. On the other hand, for the non-steady state, if ρ_{ij} or S_i or both become functions of time the coefficients may not be factored out. Even if it were possible to transform the right hand sides of the members of this system of equations, it is doubtful, in general, if a linear system would result. Furthermore, if the transformed system could be solved for the α's, their ultimate inversion would become a major mathematical problem. In spite of these difficulties there are a few favorable cases. The two-compartment system may be analytically solved. In certain instances an approximation method may be used for three-compartment systems. In most other instances we must resort to numerical solutions performed with a high-speed digital computer. Machines and methods for this are evolving rapidly. The approach in this chapter should continue to be valid in its general basic principles even if details become obsolete.

124

The Two-Compartment System

A solution for a two compartment closed system has been obtained (54). The differential equations are*

$$\frac{da_1}{dt} = \frac{\rho_{12}}{S_1}(a_2 - a_1)$$

$$\frac{da_2}{dt} = \frac{\rho_{21}}{S_2}(a_1 - a_2) \tag{1}$$

Here we may subtract one equation from the other and solve for

$$\Delta_{21} = a_2 - a_1$$

thus obtaining

$$\frac{d\Delta_{21}}{dt} = -\left(\frac{\rho_{12}}{S_1} + \frac{\rho_{21}}{S_2}\right)\Delta_{21} \tag{2}$$

For $a_1 = a(0)$, $a_2 = 0$, this equation has the following straightforward solution:

$$\Delta_{21} = (a_2 - a_1) = -a(0)\exp\left[-\int_0^t \left(\frac{\rho_{12}}{S_1} + \frac{\rho_{21}}{S_2}\right)dt\right] \tag{3}$$

For a closed system we also have the relation that the total label in the system is constant. Thus

$$S_1(t)a_1(t) + S_2(t)a_2(t) = S_1(0)a(0) \tag{4}$$

Here, of course, compartment sizes may vary with time. Solving the two equations, we obtain

$$a_1(t) = \frac{a(0)}{S}\left\{S_1(0) + S_2(t)\exp\left[-\int_0^t\left(\frac{\rho_{12}}{S_1} + \frac{\rho_{21}}{S_2}\right)dt\right]\right\}$$

$$a_2(t) = \frac{a(0)}{S}\left\{S_1(0) - S_1(t)\exp\left[-\int_0^t\left(\frac{\rho_{12}}{S_1} + \frac{\rho_{21}}{S_2}\right)dt\right]\right\} \tag{5}$$

We note that, since

$$\frac{dS_1}{dt} = \rho_{12} - \rho_{21}$$

it follows that

$$\frac{\rho_{12}}{S_1} + \frac{\rho_{21}}{S_2} = \frac{\rho_{21}S}{S_1S_2} + \left(\frac{1}{S_1}\right)\left(\frac{dS_1}{dt}\right)$$

* It may be noted that, initially, when $a_2 = 0$, the rate ρ_{12} affects the *fractional* rate of specific activity change in compartment 1 by the diluting effect of non-labeled material flowing into it. The rate ρ_{21} does not enter into the equation at all, and, if ρ_{12} were zero, a_1 would remain constant, irrespective of ρ_{21}.

Therefore, since

$$\exp\left\{-\int_0^t\left[\left(\frac{1}{S_1}\right)\left(\frac{dS_1}{dt}\right)\right]dt\right\} = \exp\left[-\int_0^t\left(\frac{dS_1}{S_1}\right)\right] = \exp\left[\log\frac{S_1(0)}{S_1(t)}\right] = \frac{S_1(0)}{S_1(t)}$$

$$a_2(t) = \frac{a(0)}{S}S_1(0)\left[1 - \exp\left(-\int_0^t\frac{\rho_{21}S}{S_1S_2}dt\right)\right] \tag{6}$$

A similar relation may be written for $a_1(t)$.

The Three-Compartment System

We may write eqs. 12 in Chapter 4 for the three-compartment system:

$$S_1\frac{da_1}{dt} = \rho_{12}(a_2 - a_1) + \rho_{13}(a_3 - a_1)$$

$$S_2\frac{da_2}{dt} = \rho_{21}(a_1 - a_2) + \rho_{23}(a_3 - a_2)$$

$$S_3\frac{da_3}{dt} = \rho_{31}(a_1 - a_3) + \rho_{32}(a_2 - a_3) \tag{7}$$

If we define Δ_{ij} as $(a_i - a_j)$, then we have

$$\frac{d\Delta_{12}}{dt} = \alpha(t)\Delta_{12} + \beta(t)\Delta_{23}$$

$$\frac{d\Delta_{23}}{dt} = \gamma(t)\Delta_{12} + \delta(t)\Delta_{23} \tag{8}$$

where

$$\alpha(t) = -(\rho_{21}/S_2 + \rho_{12}/S_1 + \rho_{13}/S_1)$$
$$\beta(t) = -\rho_{13}/S_1 + \rho_{23}/S_2$$
$$\gamma(t) = (\rho_{21}/S_2 - \rho_{31}/S_3)$$
$$\delta(t) = -(\rho_{23}/S_2 + \rho_{32}/S_3 + \rho_{31}/S_3)$$

For the case where only compartment 1 is initially labeled with specific activity $a(0)$ we can make the substitution

$$\Delta_{12} = a(0)\exp -\int_0^t\Phi(t)\,dt$$

yielding $\Delta_{23} = \Delta_{12}\psi(t)$, where

$$\frac{d\psi}{dt} = -\beta\psi^2 + (\delta - \alpha)\psi + \gamma$$

and

$$-\Phi = \alpha + \beta\psi \tag{9}$$

The differential equation in ψ is of the generalized Riccati type and is not readily solved by standard methods. In some cases Δ_{23} may be small, enabling the term in ψ^2 to be dropped as an approximation. The resulting linear equation can then be solved by quadratures.

Successive Approximations and Numerical Solutions

A good deal of theoretical knowledge concerning linear differential equations is based on solutions by successive approximations (41). The method is the basis for obtaining general analytical solutions for first-order systems of linear equations by the method of matrizants (55). Since these solutions involve infinite series of matrices they will usually be of little use in solving tracer kinetic problems. For this reason we will emphasize numerical methods in this chapter using successive approximations. Consider a first-order linear differential equation

$$dy/dx = f(x, y) \tag{10}$$

In order to solve the equation, we must obtain a value of y as a function of x which, when inserted in $f(x, y)$, causes the two sides to be equal. To obtain a unique solution, we must also specify the boundary condition. This requires the specification of the initial value $y(a)$ for the initial value $x = a$ of the independent variable. Given $y(a)$ the first step would be to try to determine an unknown value $y(b)$ for a neighboring value of the independent variable $x = b$. Once known, the result could be proved by inserting $y(b)$ into the equation and verifying that

$$y(b) = y(a) + \int_a^b f(x, y)\, dx \tag{11}$$

In the method of successive approximations, in order to obtain the correct value of $y(b)$ we may start with any initial guessed value and insert it into $f(x, y)$. The new value of $y(b)$ thus obtained by integrating the right hand side can be shown to be closer to the true value, and so it may in turn be inserted into $f(y)$ and a new integration performed. In solving equations numerically this iterative procedure is repeated by numerical integration until successive values of $y(b)$ do not differ significantly from one another at the desired level of precision of the computation.

After obtaining $y(b)$, we can use this value as a starting point for a new calculation of y at some new value of the independent variable. We may thus proceed by steps to determine y as a function of x over any desired range from the initial values $x = a$ and $y = y(a)$. The precision of the calculation will depend on the precision of the method used for numerical integration, which will in turn depend on the size of the interval between a and b. For discussions of this and other factors which affect the accuracy, reference must be made to special texts on the subject.

The method used for systems of equations is similar and can be illustrated in the case of two equations. Here we have

$$y(b) = y(a) + \int_a^b f(x, y, z)\, dx$$

$$z(c) = z(d) + \int_c^d g(x, y, z)\, dx \qquad (12)$$

We begin the calculation with guessed values of y and z, inserting them into both $f(x, y, z)$ and $g(x, y, z)$. After integration the new values are again inserted, and further iterations obtained.

Practical numerical solutions.   It is customary in many treatises on numerical integration to denote intervals of integration by the symbol h. In obtaining numerical integrals, we achieve a good compromise between labor and precision by Simpson's rule, which approximates an integral by the relation

$$y(a + 2h) = y(a) + \int_a^{a+2h} f(x)\, dx$$

$$= y(a) + (h/3)[f(a) + 4f(a + h) + f(a + 2h)] \qquad (13)$$

This method of numerical integration can be used in the numerical solution of differential equations but for the computation of y it requires that values of y be known at two values of x corresponding to h and $2h$. It may thus be used after the solution has been obtained for two points. Prior to this situation a special method is employed to perform the "starting calculation" using the relations

$$y(a + h/2) = y(a) + (h/24)[5f(a) + 8f(a + h/2) - f(a + h)]$$

$$y(a + h) = y(a) + (h/6)[f(a) + 4f(a + h/2) + f(a + h)] \qquad (14)$$

The second relation is merely Simpson's rule taken for half an h interval. Derivation of the first relation has been given by Ford (41).

Programming the numerical solution method for machine computation with the IBM 650 computer.*   Since the services of

* The work described here was performed with the assistance of the Chicago office of the Service Bureau Corporation.

high-speed, stored-program, digital computers are becoming readily available for solving equations, it seems appropriate to include a description of the numerical solution of eqs. 7 by the use of an IBM 650 Magnetic Drum Data Processing Machine (56)(57). The fact that systems of first-order equations may converge more slowly under numerical methods may be somewhat less serious when high-speed electronic computers are available. At the same time the preservation of symmetry is of some advantage. Thus the solution is performed at the level of eqs. 7 rather than at the level of the more reduced form of the equations. Although the method has been employed for only the three-compartment system, the generalization to more complex systems is obvious.

The IBM 650 machine, like other modern electronic computers of moderate to high speed, performs its calculations on data which enter its "memory" and are then processed through a series of program steps. These steps are directed by a program, which is read into another part of the memory prior to the calculation. In the case of the IBM 650, the information is read into the machine from standard IBM punched cards, which can carry up to eighty decimal digits punched in columns on the cards. The program contains a series of orders contained in ten digit numbers or "words." These words, together with other numerical information that is either being taken into the memory for processing or being stored as part of the computation, are carried in the memory up to a maximum of 2000 possible addressable storage positions. (The present discussion will apply to a standard 2000-word machine.) Data are stored as magnetized spots on the surface of a magnetic drum, and the addressable locations are situated in a series of bands along the drum. Addresses are numerical; they relate to forty bands on the drum and within each band to fifty locations. During or at the end of each calculation the results are punched out on IBM cards which can then be listed in printed form on paper or subjected to further calculation.

The program is in numerical form and instructs the machine in the performance of computations on numbers which are read into the memory before each computation. Numbers up to ten digits in length are read. Each number read or stored includes a specification of sign, either plus or minus. Reading and punching operations are performed on an entire card at one time. For each card a maximum of eight ten-digit numbers may be read in or punched. Only certain address zones in each band are available for reading, so data may enter only through addresses in these zones. The addresses that serve for entry are those whose last two digits lie between 01 and 10, or between 51 and 50. Of course, after entering, a number may be redistributed at will by suitable program orders. Punching is similarly restricted to addresses in zones 27 through 36 or 77 through 86.

A read or punch order, specifying an address anywhere in the zone, will cause all addresses in the zone to receive or emit. Since each card contains a maximum of eight numbers in the standard machine, so far as cards are concerned, the last two addresses will emit or receive zeros.

When the machine is set up for a computation, each ten-digit word in the program is stored in the memory at some location given by an address of four digits. When, during the operation of the machine, it is directed to that address, it will find an instruction contained in the first two digits of the word. For example, 71 means "punch." Following these digits, there is a four-digit address to which the instruction is related. Thus, for example, the digits 710027 mean "punch on an output card the numbers located between 0027 and 0034 in the memory." (The order 710029 would produce the same result.) The final four digits contain the address of the next instruction in the program.

The IBM system also handles alphabetic data and special characters as well as numbers. Numbers are punched as single digits. Alphabetic data are represented on cards by a system of double punches. Special characters such as "," and "*" are represented by triple punches. In preparing a program for a mathematical calculation a special system has been devised for the IBM 650 called the FOR TRANSIT system (58). This system simplifies the mechanics of programming since it permits the program to be punched on cards in a series of alphabetic statements in the "FORTRAN" language, which closely resembles the language of mathematics. The language also employs some of the special characters of mathematics. The series of statements is processed by machine to yield a program in numerical machine language. The processing of a FOR TRANSIT program includes the conversion of general orders such as + (add) or / (divide) into the actual series of machine operations required. Also the laborious searching for unoccupied memory locations and other lengthy clerical work is largely taken over by the machine.

The quantities involved in a calculation are represented by alphabetic symbols in the form of one or more capital letters. Thus in the three-compartment kinetic calculation the specific activity is A. In place of subscripted index numbers the index is appended in parentheses; thus the specific activity in compartment 1 is A(1). Since Greek letters are not included, they may be spelled out, provided not more than five letters are used. Thus RHO represents a transport rate.

Certain FORTRAN statements are not immediately obvious. For example, DIMENSION A(3) is a request to reserve space in the memory for A(1), A(2), and A(3). GO TO 20 is an instruction for the computation to avoid the next statement and go to statement 0020 instead. Statements are executed in the order in which they occur, so they need not be numbered. However, statements such as those referred to in a GO TO instruction must be identified, since the new statement to which the program is directed must be specified. The number precedes the statement and, if the statement is continued for more than one line, the statement number is followed by a line identification number. Statement numbers need not be in numerical order.

Often it may be necessary to request a calculation to be repeated in an identical manner for several index values. An example is the summing of several terms of a series. This operation is achieved by a DO order. Thus, DO 20 J $= 1$, 3 is an instruction to perform all of the statements which follow the DO statement up to and including statement 20, starting with a value of 1 for index J and ending with 3. In the summation of a series of values of X, the values $X(1)$, $X(2)$, and $X(3)$ can be successively entered by addition into an initially empty register. The only requirement is to precede the entry order with a DO instruction.

Since the power of modern computers lies in their ability to perform primitive "decision" operations, the list of permissible statements includes the statement IF. Thus IF(X) 5, 10, 7 means that the next statement to be performed is number 5, 10, or 7, depending upon whether X is less than, equal to, or greater than zero.

The FORTRAN language is particularly simple in arithmetical instructions which are written out in the usual manner. Thus

$$(A + B * C)/D$$

means add A to the product of B and C and divide the result by D. As in ordinary mathematics parentheses are used to specify the order of arithmetical operations.

FORTRAN statements for three-compartment solution.   Table 3 shows the statements for the solution of eqs. 7. The format is that of FOR TRANSIT I(S) which applies to machines having special character devices. Modification to the basic alphabetic 650 or, to 1620 FORTRAN, is not difficult. The variables involved and their designations are given in Table 4. The four-digit statement numbers are at the extreme left and occupy columns 2 through 5. In their absence zeros are punched. Line numbers are blank except in the case of more than one line. The first "continuation card," which continues the statement, is numbered 1, the second 2, and so on up to 9. These numbers are punched in column 6. Columns 7 through 36 contain the statements. As indicated in the table, not every column need be punched. The program begins with a DImENSION instruction which assigns memory locations for all "subscripted" variables involved in the calculations. This is carried over on

Table 3

FORTRAN statements used for preparing IBM 650 program for solution of differential-difference equations for a non-steady state three-compartment system

```
C       FORTRANSIT PROGRAM FOR SOLUTION OF
C          THREE COMPARTMENT SYSTEMS

0000    DIMENSION RHO(3,6), S(3,3),
00001   A(3,3), T(3), PHI(3,3),
00002   A1(3,3)
0001    READ 1,A(1,1), A(1,2), A(1,3)
0000    K = 1
0002    DO 3 J = K,3
0000    READ 1, RHO(J,1), RHO (J,2),
00001   RHO(J,3), RHO(J,4),
00002   RHO(J,5), RHO(J,6)
0003    READ 1, S(J,1), S(J,2),
00031   S(J,3), T(J),  D
0000    IF(D) 31, 31, 53
0053    TCK = D
0000    IF (T(1)) 30, 4, 30
0004    IF (T(2) - D/2.) 30, 5, 30
0005    IF (T(3) - D) 30, 6, 30
0006    L = 1
0007    DO 12 J = 1,3
0008    PHI(J,1)=(RHO(J,1)*(A(J,2)-A
00081   (J,1))+RHO(J,2)*(A(J,3)-A(J,1
00082   ))))/S(J,1)
0000    PHI(J,2) = (RHO(J,3)*(A(J,1)-
00001   A(J,2)) + RHO(J,4)*(A(J,3)-
00002   A(J,2))))/S(J,2)
0000    PHI(J,3) = (RHO(J,5)*(A(J,2)-
00001   A(J,3)) + RHO(J,6)*(A(J,1)-A
00002   (J,3))))/S(J,3)
0000    IF (L - 3) 10, 13, 12
0010    L = L + 1
0000    DO 11 I = 1,3
0011    A(J+1, I) = A(J,I) + PHI(J,I)
00111   *D/2.
0012    CONTINUE
0013    DO 15 I = 1,3
0014    A1(2,I) = A(1,I) + D*(5.*PHI
00141   (1,I) + 8.0*PHI(2,I) - PHI
00142   (3,I))/24.
0015    A1(3,I) = A(1,I) + D*(PHI(1,I
00151   ) + 4.0*PHI(2,I)+PHI(3,I))/6.
0000    DO 18 I = 1,3
0016    IF (.00001 - ABSF(A1(2,I) -
00161   A(2,I))) 27, 27, 17
0017    IF (.00001 - ABSF(A1(3,I) -
00171   A(3,I))) 27, 27, 18
0018    CONTINUE
0000    DO 23 J = 1,3
0022    PUNCH 1, RHO(J,1), RHO(J,2),
00221   RHO(J,3), RHO(J,4),
00222   RHO (J,5), RHO(J,6)
0023    PUNCH 1, S(J,1), S(J,2),
```

132

Table 3

(*Continued*)

```
00231  S(J,3),  A(J,1),
00232  A(J,2),  A(J,3),  T(J)
0000   DO 25 I = 1,3
0024   PHI(2,I) = PHI(3,I)
0025   A(2,I) = A(3,I)
0000   D1 = D
0000   K = 3
0000   GO TO 2
0027   DO 29 I = 1,3
0028   A(2,I) = A1(2,I)
0029   A(3,I) = A1(3,I)
0000   L = L + 1
0000   GO TO 7
0030   PAUSE 0030
0000   GO TO 6
0031   TCK = TCK + D1
0000   IF (T(3) - TCK) 51, 32, 51
0032   DO 33 I = 1,3
0033   A(3,I) = A(2,I) + D1*PHI(2,I)
0034   PHI(3,1) = (RHO(3,1)*(A(3,2)-
00341  A(3,1)) + RHO(3,2)*(A(3,3)-
00342  A(3,1)))/S(3,1)
0000   PHI(3,2) = (RHO(3,3)*(A(3,1)-
00001  A(3,2)) + RHO(3,4)*(A(3,3)-
00002  A(3,2)))/S(3,2)
0000   PHI(3,3) = (RHO(3,5)*(A(3,2)-
00001  A(3,3)) + RHO(3,6)*(A(3,1) -
00002  A(3,3)))/S(3,3)
0000   DO 38 I = 1,3
0038   A1(3,I) = A(1,I) + D1*(PHI(1,
00381  I) + 4.*PHI(2,I)+PHI(3,I))/3.
0000   DO 41 I = 1,3
0040   IF (.00001 - ABSF(A1(3,I) -
00401  A(3,I))) 48, 48, 41
0041   CONTINUE
0042   PUNCH 1, RHO(3,1), RHO(3,2),
00421  RHO(3,3), RHO(3,4), RHO(3,5),
00422  RHO(3,6)
0043   PUNCH 1, S(3,1),S(3,2),
00431  S(3,3), A(3,1),
00432  A(3,2), A(3,3), T(3)
0000   DO 47 J = 1,2
0044   DO 46 I = 1,3
0045   A(J,I) = A(J+1,I)
0046   PHI(J,I) = PHI(J+1,I)
0047   CONTINUE
0000   GO TO 2
0048   DO 49 I = 1,3
0049   A(3,I) = A1(3,I)
0000   GO TO 34
0051   PAUSE 0051
0000   GO TO 32
0000   END
```

Table 4

Variables used in the FOR TRANSIT program and their designations

D, D1 = Time increment corresponding to h in ordinary numerical integration.

T(J) = Jth time value where T(1) = initial time, T(2) = T(1) + D/2, T(3) = T(1) + D.

A(J, I) = Specific activity in compartment I at time T(J).

S(J, I) = amount of S in compartment I at time T(J).

RHO(J, I) = Ith transport rate at time T(J),

where I = 1 denotes transport from compartment 2 to 1,

I = 2 denotes transport from compartment 3 to 1,

I = 3 denotes transport from compartment 1 to 2,

I = 4 denotes transport from compartment 3 to 2,

I = 5 denotes transport from compartment 2 to 3,

I = 6 denotes transport from compartment 1 to 3.

PHI(J, I) = dA/dt in compartment I at time T(J).

A1(J, I) = Approximate values of A(J, I) obtained during the course of the iterative calculations.

TCK = time carried in the calculation and tested against time on the input cards. This provides a check against cards getting out of order.

two additional continuation cards. Statement 0001 is an instruction to read the A data from a set of input cards.

The "starting calculation" based on eqs. 14 is signaled by the presence of a non-zero value of D on the cards. This is tested in the twelfth line of the program where, in an unnumbered IF statement, the program proceeds directly to the following one, number 0053, if D > 0. This part of the program represents the starting calculation. If D = 0, the program proceeds on to 0031, which begins the "continuing calculation." If the starting calculation is to be performed, then three sets of A, RHO, S, and T—corresponding to J = 1, 2, and 3—must be read. We achieve this by setting up an index K, equating it to 1, and placing the read order under a DO instruction. This causes values to be read for J = 1, 2, and 3. In the later continuing part, only one set of data for J = 3 is required. This is managed in an unnumbered statement 2 lines below 0025, where there is now made the assertion "K = 3," which modifies the read instruction in preparation for its subsequent re-use.

As we start the calculation, instruction 0053 causes TCK to be set up as a check value of the time. The initial value D for TCK assumes that all calculations start with an initial zero time value. The next series of instructions insures that the time values punched on the cards agree with the increment D. If this test fails, the next statement referred to is number

0030, and the machine stops and awaits the action of the operator. Instruction 0006 sets up an index L which starts with the value 1. This index permits the program to compute initial "guess" values of A before it begins the series of iterations. The guess values require the computation of the derivatives PHI, which begin with the initial derivative at T = 0, evaluated from initial RHO and A values for all three compartments by means of a DO instruction extending from instruction 0007 through 0012. This calculation causes the program to perform step 0008 through 0011 three times—once for each value of J. On the first performance of the sequence the initial derivatives for each compartment are computed.

Since L = 1, the program, upon arriving at the IF statement, will continue directly to the next statement, 0010. The value of L will then increase to 2, and the "guess" values for the first half interval A(2, I) will be computed by assuming a simple linear increase of the A's at rates given by the initial derivatives. Since the computation is included in a second DO statement, the calculation will be made for all three values of I representing all three compartments. After the next cycle in which the A(3, I)'s are "guessed," the final cycle, yielding the derivatives for the third time value, is completed.

At the IF statement, since L now equals 3, the program transfers to statement 0013. Here the iterations begin for computation of the A(2, I)'s and the A(3, I)'s from the initial values of A and the derivatives PHI at all three points for each compartment. On each cycle the improved values A1 are obtained from the previous A values by the arithmetical operations set forth in statements 0014 through 0015. Computations for each compartment are performed by including the statements within a DO instruction with index I ranging from 1 to 3.

When statement 0015 is completed, a series of tests are made to determine if the A1's differ from the A's by more than 10^{-5}. In each test the absolute value (ABSF) of (A1 − A) is subtracted from 0.00001 and the sign of the result examined. If any of the tests fail, the program is directed to statement 0027, where all A's are replaced by A1's. Then a new cycle occurs involving both a new computation of the PHI's and an evaluation of a new set of A1's.

Once the set of IF tests is passed completely, the starting calculation is complete. Then, beginning at statement 0022 punching begins. The values of A for all compartments are punched for zero time, for time equal to half a D interval, and for a full interval. For additional convenience, the T values and the original RHO and S data are also punched. At statements 0024 and 0025, the PHI and A variables are redefined in terms of a full D increment. Since D is to be read from the cards which will now be blank at that location, it must be retained in the memory as D1. The program is now directed to statement 2 and a new set of cards is read, but, since K = 3 now, only the values for J = 3 will be read.

Since D is now blank, it will appear in the memory as zero. As a result the IF test will direct the program to statement 0031 for a continuing calculation. A check test follows 0031, to insure that the time value just read is correct; then a series of "guess" values for A(3, I) is computed. For the first continuing calculation, these values will be for T = 2D. Now these values, plus the A(1, I) for T = 0 and the A(2, I) for T = D, already computed and retained, are used to compute the corresponding derivatives PHI in statement 0034 and in the two unnumbered statements which follow 0034. Statement 38 completes the computation of the improved values A1. A test is then made at statement 0040 which, if satisfied, completes the computation of the A's with punching out in statement 0042 and 0043. Following this computation in statements 0045 and 0046, the variables are redefined for a new calculation one D interval higher. The END statement must appear at the conclusion of all FOR TRANSIT programs.

The FOR TRANSIT program is processed by the usual procedures described in the IBM manual (58). In the first phase of processing, it will be necessary to precede all of the statements in Table 3 by a function-title card for the absolute-value calculation "ABSF." Columns 1 through 30 of this card will contain

$$0000001500 \quad 0061628266 \quad 9294657200$$

and, in addition, "12" punches will appear in columns 2, 10, 20, and 30.

The subroutine ABSF, which converts a number to its absolute value, consists of two cards containing the instructions in symbolic form. They are inserted beneath the first (REG) card of the series of symbolic instruction cards to be processed through phase 3. The instructions and their appropriate columns are

columns	43–47	48–50	51–55	57–61
card 1	EOOAY	STD	ABSFE	
card 2		RAM	8002	ABSFE

The blank spaces are left unpunched and the system automatically provides addresses during phase 3 processing into the final machine-language program.

Input and output format.   The data cards which are processed by the program must contain the initial specific activities A(0) and the various values of RHO and S for the various values of the time. Each value of time T must also be given. All input values are given in "floating-point" notation in which the first eight digits punched on the card are the first eight non-zero digits of the number. The last two digits give the floating-point index.

This index is the power of ten by which the number is to be multiplied in order to shift the decimal to its proper position. Since decimal shifts may proceed either way, the index is increased by 50 in its final expression. Thus, if a shift of 3 points to the left is required, the index will be $50 - 3 = 47$. For a shift three points to the right, it will be 53. In the floating-point system the decimal before shifting is located at the left of the first digit. Here is an illustration of the system: if the floating-point number is 2345000048, the actual number is 0.0023450000. The sign for every number is punched over the extreme right-hand digit. The FOR TRANSIT system employs a console control which can be set to stop the calculation automatically, if numbers involved in arithmetic operations fall outside the range 10^{-51} to 10^{49}. This feature should be utilized in the calculations.

Card layout. The layout scheme for the first nine cards containing input data is shown in Table 5. Each line represents a consecutive card

Table 5

Input card layout for solution of differential-difference equations. Zeros must be punched in blank spaces, and a "12" punch must appear in Column 73.

			Field			
Card	1	2	3	4	5	6	7
1.	A(1, 1)	A(1, 2)	A(1, 3)	——	——	——	——
2.	RHO(1, 1)	RHO(1, 2)	RHO(1, 3)	RHO(1, 4)	RHO(1, 5)	RHO(1, 6)	——
3.	S(1, 1)	S(1, 2)	S(1, 3)	T(1)	——	——	——
4.	RHO(2, 1)	RHO(2, 2)	RHO(2, 3)	RHO(2, 4)	RHO(2, 5)	RHO(2, 6)	——
5.	S(2, 1)	S(2, 2)	S(2, 3)	T(2)	——	——	——
6.	RHO(3, 1)	RHO(3, 2)	RHO(3, 3)	RHO(3, 4)	RHO(3, 5)	RHO(3, 6)	——
7.	S(3, 1)	S(3, 2)	S(3, 3)	T(3)	D	——	——
	end of starting group						
8.	RHO(3, 1)	RHO(3, 2)	RHO(3, 3)	RHO(3, 4)	RHO(3, 5)	RHO(3, 6)	——
9.	S(3, 1)	S(3, 2)	S(3, 3)	T(3)	D = O	——	——

which, in turn, contains eighty columns. The location of the hole in a given column specifies the digit which is to appear in the column. Digits from 0 to 9 can be punched, and, in addition, there is room for 11, or "X," and 12, or "Y," punches, which are used for minus or plus signs respectively. They may also be used for control purposes, and in the FOR TRANSIT system all cards must have a control Y punch in column 73. Cards are divided into seven fields with the eighth reserved for card identification purposes. The input numbers are punched in these fields, seven to a card. When a field does not contain data, zeros must be punched. All ten digits in a field are used for identification of the variable, and the sign is punched over the tenth.

The first seven cards form the group which starts the computation. The first card of the group contains the initial values of specific activity A(1), A(2), and A(3), corresponding to compartments 1, 2, and 3 respectively. On the next card are the RHO values and finally on the third card the S values and the time. In starting the solutions it is recalled that an initial value T(1) of the time (usually zero) is required. Following this initial value is a value T(2), corresponding to a half-interval value $T = D/2$, and finally a value T(3), corresponding to a full interval $T = D$. The first index for RHO and S is identical with the time index and the second represents the compartment. Thus S(2, 1) represents the amount of S in compartment 1 at time T(2). For the second index the following relation holds between the FORTRAN language and the previous notation of this volume:

FORTRAN Index	Equations 7
1	ρ_{12}
2	ρ_{13}
3	ρ_{21}
4	ρ_{23}
5	ρ_{32}
6	ρ_{31}

After the starting calculation is completed, the continuing calculations are performed serially, and cards are read in groups of two—groups whose representation is identical with the last two starting cards, but with $D = 0$.

Table 6

Output card layout for differential difference equations

Card	1	2	3	4	5	6	7
				Field			
1.	RHO(1, 1)	RHO(1, 2)	RHO(1, 3)	RHO(1, 4)	RHO(1, 5)	RHO(1, 6)	———
2.	S(1, 1)	S(1, 2)	S(1, 3)	A(1, 1)	A(1, 2)	A(1, 3)	T(1)
3.	RHO(2, 1)	RHO(2, 2)	RHO(2, 3)	RHO(2, 4)	RHO(2, 5)	RHO(2, 6)	———
4.	S(2, 1)	S(2, 2)	S(2, 3)	A(2, 1)	A(2, 2)	A(2, 3)	T(2)
5.	RHO(3, 1)	RHO(3, 2)	RHO(3, 3)	RHO(3, 4)	RHO(3, 5)	RHO(3, 6)	———
6.	S(3, 1)	S(3, 2)	S(3, 3)	A(3, 1)	A(3, 2)	A(3, 3)	T(3)
	end of starting group						
7.	RHO(3, 1)	RHO(3, 2)	RHO(3, 3)	RHO(3, 4)	RHO(3, 5)	RHO(3, 6)	———
8.	S(3, 1)	S(3, 2)	S(3, 3)	A(3, 1)	A(3, 2)	A(3, 3)	T(3)

Thus the time will always be represented as T(3) in the continuing calculation. The value of D is punched in columns 41 through 50 in the last card of the starting group only. Table 6 gives the punching layout of the output data. The representation is similar to Table 5.

Preparation of input data—use of 602A calculating punch. Of course the solutions which are obtained in a given calculation will depend on the numerical values punched on the input cards. These data may be worked up by hand and punched manually on the cards in simpler instances. Care must be taken that S values are not too small, since in this case the right hand side of one or more equations may become too large for the machine to accommodate. An overflow will then be signaled, and the computation will terminate.

In some instances the calculation of input data becomes laborious and machine methods may be preferred. Often the preliminary computation will be so simple that it can be performed with less-complex IBM equipment. In many urban centers, the services of the IBM 602A calculating punch are made available on an hourly charge basis by the Service Bureau Corporation. This machine is slow and is much more limited in memory capacity and in program size than is the IBM 650. However, simple computations are fairly readily programmed by appropriate connections of its wired control panel, and the machine is inexpensive to operate.

The first step in the preparation of input data is the calculation of the ρ values. A typical wiring diagram for computing ρ by the summation of polynomials is shown in Fig. 40. The planning chart layout is shown in Fig. 41. The computation involves the evaluation of

$$\rho = a_0 + a_1 t + a_2 t^2 + a_3 t^3 + a_4 t^4$$

This calculation is done by forming the product $a_4 t$, adding a_3, multiplying by t, and repeating the process. The value of t used in the computation is punched by the machine, and, when the calculation is completed, ρ is also punched. The layout of the card is shown in Fig. 42, along with the layout of the input card. This card carries the a's, the initial value of the time, and the value of the time increment D. The card is identified by an "11" or "X" punch in column 80.

Cards fed into the machine start with the input card followed by a series of cards into which the serial values of ρ and t are to be punched. These cards are initially gang punched with the appropriate values of both the solution number "S.N.," which identifies the particular calculation, and the index, which identifies the particular ρ among the total of six which are to be computed for each solution. This index is the same as that later

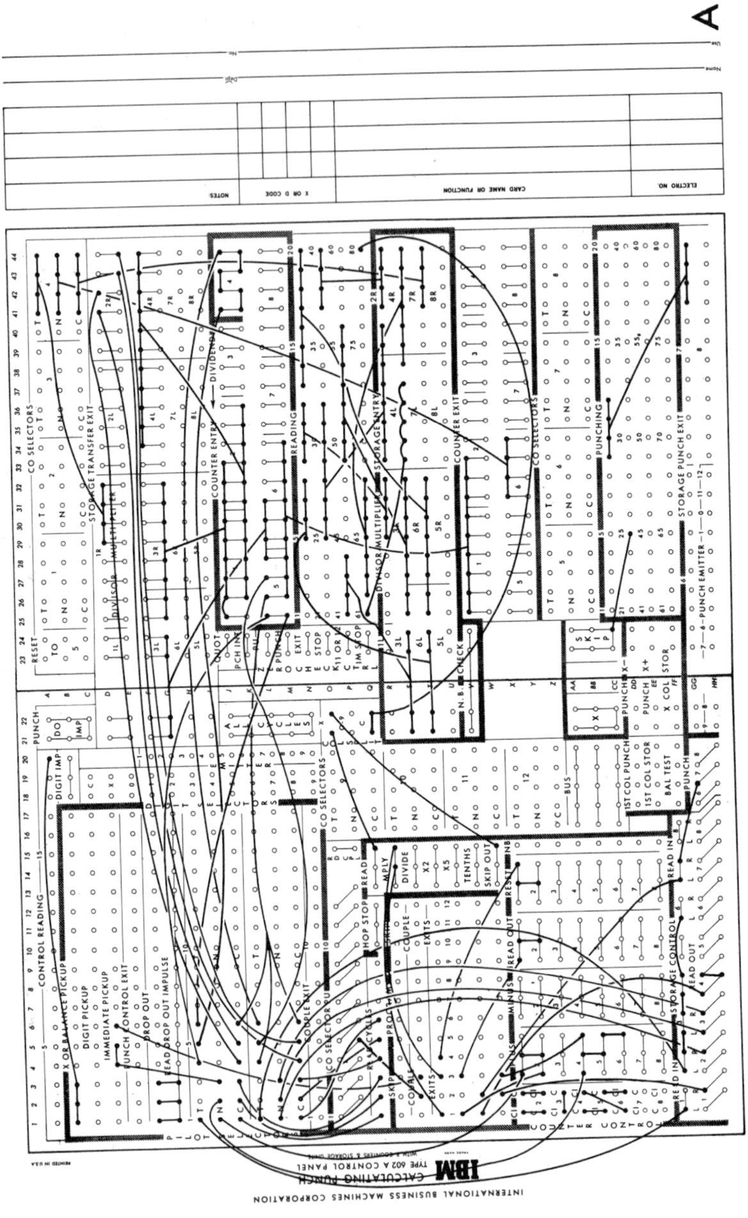
IBM
TYPE 602 A CONTROL PANEL
INTERNATIONAL BUSINESS MACHINES CORPORATION

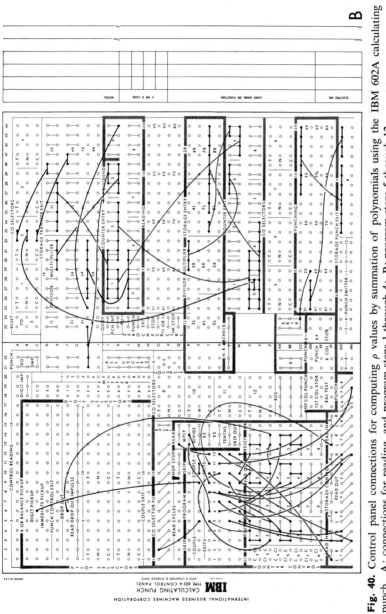

Fig. 40. Control panel connections for computing ρ values by summation of polynomials using the IBM 602A calculating punch. A: connections for reading, and program steps 1 through 4; B: program steps 5 through 12.

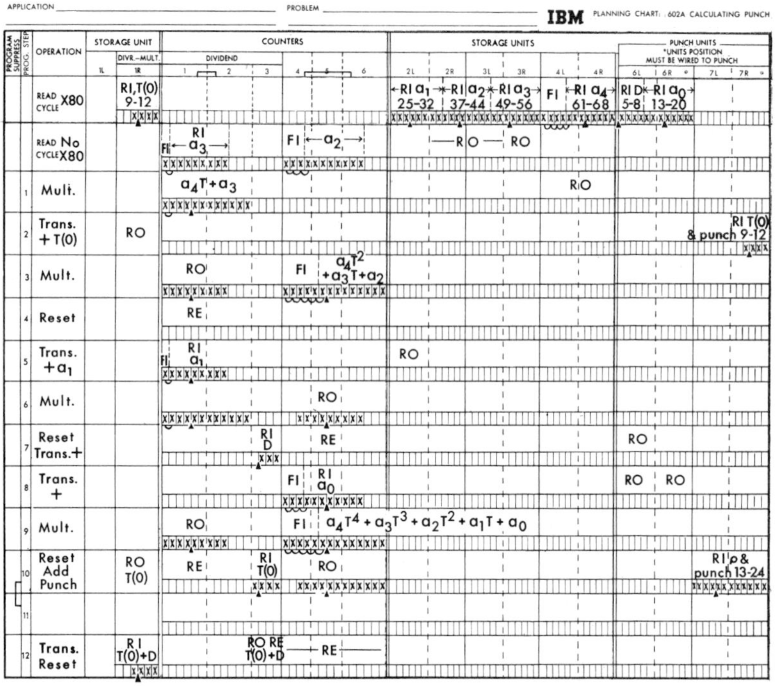

Fig. 41. Planning chart for ρ calculation. Notation is similar to common usage for IBM machines, RI = "read in," RO = "read out," RE = "reset," and Trans. = "transfer." RI in the counters is executed by impulsing +. Note that in some instances as in storage 4L, top line, digits are connected together to fill in nines to the left of the highest order digit in case the number becomes complemented. This is indicated by the designation FI. Line 1 represents the operations to be executed on reading a card containing an X-punch in column 80. Line 2 represents the alternative case where no X-punch occurs. Following the read cycle, the program cycles 1 through 12 are executed.

used in the FORTRAN program. Also included in the prepunched information is the time increment D.

When each ρ value is completed, the increment D is added to the current value of t and the new value, which replaces the old in the machine storage, is used in the succeeding calculation. The a's remain unaltered in the machine for the entire calculation corresponding to a given S.N. It will be noted in Fig. 42 that a total of 12 digits are allotted for ρ. The reason for this is that the 602A uses "fixed point" arithmetic in which the decimal point is unvarying. Therefore, to allow for a large range in values, a larger space must be provided.

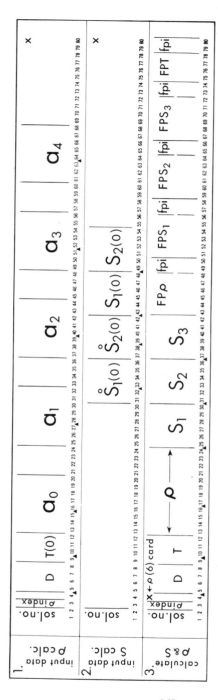

Fig. 42. Card layouts for 602A calculations. All 80 card columns are shown. Decimal points are represented by small triangles. Solution number is punched in columns 1 and 2 to identify the calculation. Layout at the top is for the introduction into the machine of *a*'s and other data required for computing ρ values. Note "ρ index," in column 3, which distinguishes the 6 different rates from one another. Notation is that of Table 4. Layout in the second line provides the initial information required for computing S_1 and S_2 by numerical integration. $\mathring{S}_1 = dS_1/dt, \mathring{S}_2 = dS_2/dt$, values being given in each case for initial time $T(0)$. Both top and middle cards have an X-punch in column 80. The results of the computations are punched according to the layout on the bottom line. The sixth card is identified by an X-punch in column 4. It carries not only ρ_6 but also the S values in columns 25 through 42. Space is also reserved on the remainder of the card for later reproducing the variables in floating point notation, including floating point indices "fpi."

143

602A CALCULATING PUNCH CONTROL PANEL
WITH 4 COUNTERS & STORAGE UNITS

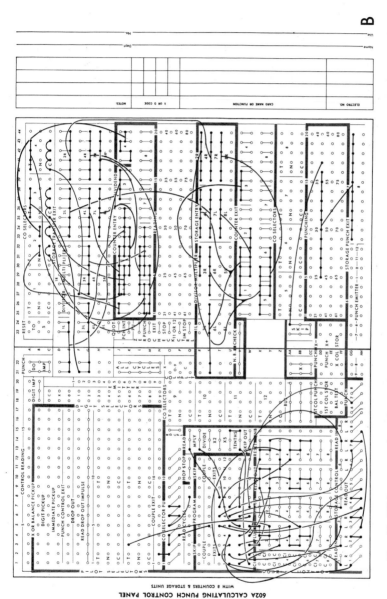

Fig. 43. Wiring connections for computing S_1 and S_2.   A: Reading instructions;   B: program steps.

Although all ρ values are positive, it is possible for a's to have negative values. If a is negative, its "nines complement," obtained by subtracting each digit of the number from 9, is punched on the card instead of the original negative value. This method of handling negative numbers requires that the total number of digits to the left of the decimal point in any of the a's must be one less than the number of assigned digit spaces on the card. In panel wiring the use of complements necessitates a common connection between the highest digits in some instances. This filling in of digits is indicated by "FI" in Fig. 41.

In addition to the ρ values the values of S must be calculated as part of the input data. Figure 43 shows the connections for computing the compartment sizes S_1 and S_2, and Fig. 44 shows the planning chart layout. From these computations it is a simple matter finally to obtain S_3 from the relation

$$S_3 = 1 - S_1 - S_2$$

The board layout for this last calculation is extremely simple. Its wiring has not been shown since it is a simple "crossfooting" operation which is familiar to all who routinely use the 602A machine. The S values are carried on the sixth ρ card (Fig. 42). For each solution number the initial values $S_1(0)$ and $S_2(0)$ as well as the initial time derivatives $\overset{\circ}{S}_1(0)$ and $\overset{\circ}{S}_2(0)$, must be entered into the machine. These values are punched on a leading card identified by an "11" or "X" punch in column 80 (Fig. 42). After this card has been entered, the six ρ cards are read in succession, and the sixth card ("ρ_6 card" in Fig. 44), identified by an "11" punch in column 4, is held in the machine. By appropriate additions and subtractions of the various ρ values as they are read, we compute the final net rates of increase or decrease of S_1 and S_2. These values are numerically integrated by the trapezoidal rule and punched in the card. Since the first integrated value corresponds to T = T(0) + D, the ρ cards for zero time are removed and replaced after the S calculation is complete. Values of $S_1(0)$ and $S_2(0)$ are punched on the sixth ρ card manually before it is returned to the deck.

This program, as well as the ρ program, is designed for a standard 602A machine which possesses seven pilot selectors and eight co-selectors. Control reading brushes 2, 8, and 20 are connected to read in card columns 3, 4, and 80, respectively. In other situations, appropriate shifts in brush connections will be required.

Final assembly of input data.   The final assembly of the input data in the format of FOR TRANSIT first requires conversion of the data from fixed point to floating point form. This change can be effected by passing the cards through a sorting machine looking for zeros in successive columns. As soon as a non-zero column is encountered, the card is separated, and all cards of its type are thereby identified as to the number of decimal point shifts required. In this way the shifted number and its floating point index can be punched in an appropriate field by standard

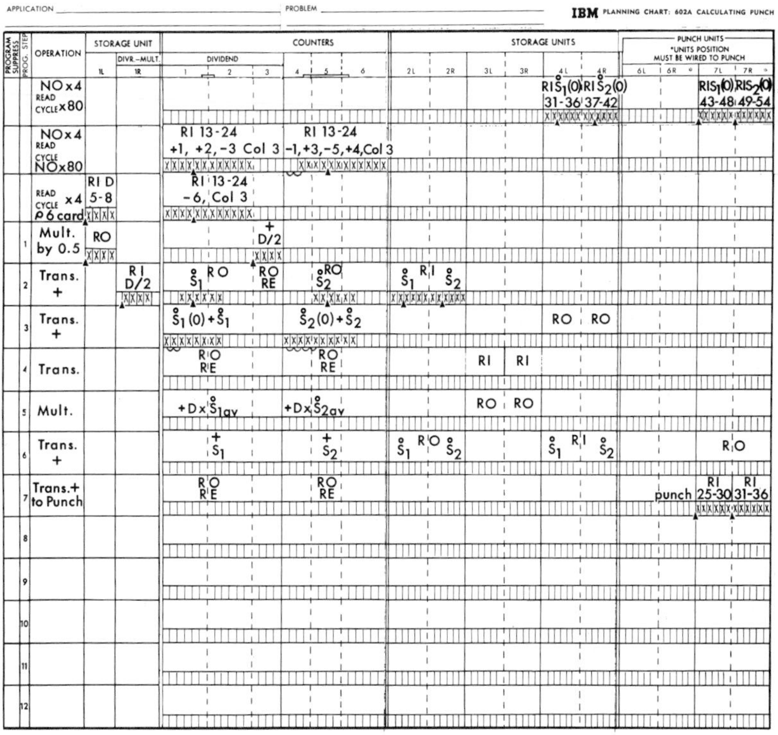

Fig. 44. Planning chart for computing S_1 and S_2. Note: instructions in lines 2 and 3 relate to the $+$ or $-$ impulse to be given to a particular bank of counters depending upon the value of the digit in column 3 of a card. For example in line 3 the instruction "RI 13–24 $-$ 6, Col 3" means "if column 3 on the card contains a six digit, read in data negatively from columns 13 to 24." Note digits filled in to the left in some instances as in Fig. 41. Line 3 represents the operations to be executed during the read cycle in the case where an X-punch occurs in column 4, i.e., "ρ_6 card."

card reproducing techniques. The final step is to reproduce the data and the identifying information on new cards to form the 650 input deck.

Typical results. Figure 45 shows a typical result of a numerical solution for a three-compartment system. The initial values and the transport rates as functions of time are also included. The size of compartment 1 does not change, but compartment 2 shrinks and compartment 3 grows as time progresses. Initially the transport between compartments 2 and 3 is a pure exchange process, but the rate from 3 to 2 decreases with time. At $t = 0.2$, the rate from 3 to 2 would reach zero, but at this point compartment 2 vanishes. Because of the large size of compartment 1 and the rapidly diminishing rate at which label is received from compartment 3, a_2 rapidly approaches a_1. Therefore, from $t = 0.2$, the system behaves as a

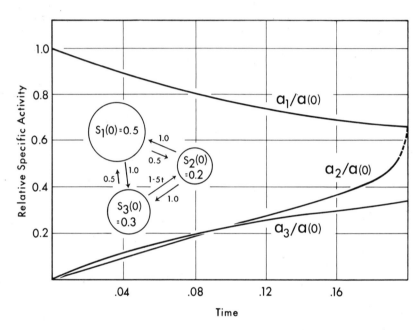

Fig. 45. Results for a typical three-compartment system where one compartment ultimately vanishes. Note rapid merging of $a_1(t)$ and $a_2(t)$ as time approaches 0.2.

two-compartment system. It may be noted that just before $t = 0.1$ the specific activities of compartments 2 and 3 cross one another.

Figure 46 shows the effect of variation in exchange rate between two compartments in a pure exchange system with constant compartment sizes. All opposing rates are equal and opposite, so there is no net amount of S accumulating or being lost from any compartment. The effect of varying exchange between compartments 1 and 2 is shown in the figure. Initially the exchange is slow; therefore compartment 1 equilibrates primarily with compartment 3. The activity in compartment 2 remains near zero. With the onset of a "bump" in the exchange rate in the region of $t = 0.25$, a_2 begins to rise rapidly, and a_1 drops more rapidly than before. As the exchange between 1 and 2 falls to nearly zero at $t = 0.5$, the a_1 and

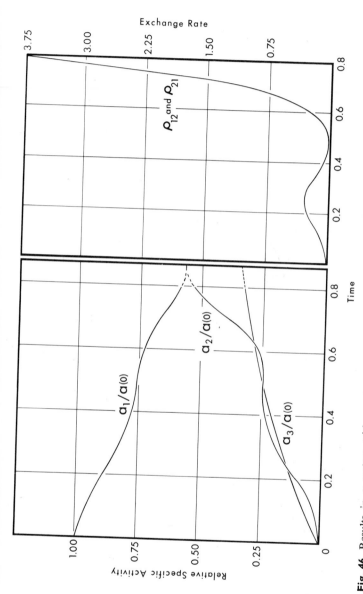

Fig. 46. Results in a system with constant compartment sizes, but variable exchange rate between compartments 1 and 2. Specific activities are shown at the left. Equal rates ρ_{12} and ρ_{21} are shown as a function of time at the right. Constant values are $\rho_{13} = \rho_{31} = 0.2$ and $\rho_{23} = \rho_{32} = 0.1$.

a_2 curves flatten. Finally, there is a large upshoot of the exchange rate above $t = 0.7$, and the two specific activities rapidly merge.

General Remarks Concerning the Use of Digital Computation

The decision as to whether or not to turn to large computing machines in solving a tracer kinetic problem will depend on a number of factors. Of course, most workers will have access to some sort of electrically driven desk calculator. If only one solution is required—for example, that of a three-compartment system—and if a few widely spaced points can be accepted, it may be advisable to resort to hand calculation with the desk machine. Of course, a systematic layout and an orderly approach will save a great amount of labor. Hints as to procedure and advice concerning the making of good initial guesses of the beginning values will be found in standard textbooks (41).

When we do turn to machine computation, the availability of a prepared program proves to be a great inducement. An ever-increasing library of programs can be found in modern computing centers, and there is always the possibility that the appropriate program may be found. Even if a program is available, however, it may not conform closely to the requirements of the user. The program in Table 3 will be of no use if the system is not a closed one. If six-figure accuracy is required, that program will be useless, since the iterative loop terminates if successive values differ by only 10^{-5}. Nevertheless, programs can often be modified fairly easily by an experienced programmer. The novice should certainly have access to the services of such an individual. This consideration is particularly important if there is a chance of a program containing errors, since the search for these errors by inexperienced workers can prove costly. Even workers with a fairly good knowledge of computing machine techniques will benefit from such assistance, particularly if the programmer is an expert with a particular data-processing system. Such individuals seem to develop an uncanny faculty for locating troubles or for anticipating them.

Two important principles will help to prevent difficulty. The first is to be as literal minded as possible and to cultivate an interest in details. For example, in the author's own experience, a "1," erroneously written in place of an "I" in the program shown in Table 3, played havoc with the calculation. The second principle is to take nothing for granted. It might be assumed that a computing machine is always right, but this condition is not necessarily true; therefore, the use of a good test calculation is of considerable value. In the present calculation, the test may consist of a system of equations with constant coefficients whose solutions may be readily obtained by other simpler methods. Another valuable checking procedure is the repetition of a calculation. Finally, a solution may be assumed to be correct if it satisfies the equations. The less-advanced types of computing equipment are not as readily programmed to detect machine errors and, of course, the problem of human error in preparing and punching input data must be taken into account. Here an experienced operator and the processing of cards through a verifier after punching can frequently save much time and money. Although the system described here uses cards, more sophisticated methods for handling data have recently appeared. Nevertheless, in the author's own laboratory, the card systems have been found to be highly convenient for present needs.

8

Non-Compartmental Systems—
Basic Tracer Equations

If we inspect the equations describing the movement of label between compartments in a multicompartment system, we are struck with the similarity in form between these and the equations of diffusion of a solute between stirred-solvent compartments separated by permeable membranes. In order to understand this analogy, we must consider the basic principles of tracer theory and proceed beyond the limits of tracer movements purely in systems of compartments. We therefore inquire into the movement of tracer substances in general systems whose composition and parameters may vary continuously, both spatially and temporally. Because of the analogy between tracer movement and diffusion, we will look for similar transport equations and recall the Fick law equation of diffusion as a possible model.

The Tracing of Substances in Continuous Systems

Much basic information can be inferred from the study of steady-state systems so our analysis will begin from this point. A system can be recognized as being in a steady state if the concentration of S is constant in time at any point. Of course, it may not be spatially constant, and the transport rates may vary from point to point and with time. If the system is in a steady state, however, the total amount of S entering a differential-volume element dv must equal the total amount leaving it. This condition will hold if all transport processes represent exchange or microscopic random motions. Of course a steady state, as we have defined it, can occur when a uniform current of S sweeps through the system carrying material into and out of all volume elements at the same rate, and a uniform system in equilibrium with constant specific activity throughout will appear to behave in this manner if it is observed from a system of moving coordinates. We will avoid this obviously ambiguous situation by postulating that the coordinates are stationary with respect to the system. The system will be considered to be surrounded by closed boundaries through which S cannot flow, and to contain no sources or sinks of S.

Since the properties of the system are to be continuously variable, it is necessary to consider its infinitesimal elements. Let us consider at point P an infinitesimal differential element of area $\mathbf{dA}$ (Fig. 47) through which S is flowing or exchanging at a certain rate. If we orient the element until the amount of label flowing through is maximal, we will define the flux $\mathbf{J}^*$

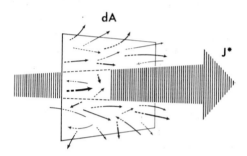

Fig. 47. Differential element of area $\mathbf{dA}$ and $\mathbf{J}^*$, the transport of labeled S through it.

at P as the amount of label traversing $\mathbf{dA}$ per second per unit area. It is immediately evident that $\mathbf{J}^*$ is a type of average quantity. Molecules of label may approach $\mathbf{dA}$ from many directions, and, if they traverse $\mathbf{dA}$, they will be included in the flux. If one wished to determine, by the tracer method, the distribution in direction of these molecules, it would be necessary to make measurements in an infinite number of infinitesimal compartments surrounding P. Such measurement is not possible by methods presently known. Similar problems would arise if we wished to determine the distribution in the molecular velocities or other detailed information of this sort.

Diffusion versus Interfusion

In a steady-state system, as we have defined it, there will be no net movement of S. Labeled species of S will mingle with the unlabeled in an approach to uniform specific activity throughout the system.† Whenever

---

† In concerning ourselves with the mingling of labeled and unlabeled species, we are reminded of the differences which Gibbs recognized between the *generic* definition of probability, in which permutations of identical particles were permitted, and the *specific* definition, in which permutations represented different realizations of the state of the system.

a molecule of S moves, on the average, another will take its place. The probability that the new molecule is a labeled one depends upon the relative fraction of S in the vicinity which is labeled, i.e., the specific activity. Thus qualitatively it may be seen that the current of labeled material depends upon the specific activity. In the process of diffusion in a stationary solvent at low concentration, the probability that the molecule of solute will replace a molecule of solvent will vary with the density of solute molecules, i.e., their concentration. We thus see that there are analogies between solvent and S and between solute and *labeled S*. In diffusion the factor controlling the speed with which solute and solvent mix is the "diffusion coefficient." Its magnitude will depend on the rate with which labeled and unlabeled species of S mingle. In the analogy of S and labeled S something more must be invoked. The name "interfusion" has been coined for this different type of process (4).

We may question whether there is any real difference in these two situations. If labeled and unlabeled S are mingling and the concentration of S is uniform throughout the system, then we may prefer to believe that labeled S is mingling with solvent molecules (if S were a solute, for example). Why, then, is it necessary to invoke anything different from simple diffusion? In the instance just cited either picture will suffice, but we may require equations which will be of broader utility. How can we describe the mixing of labeled and unlabeled pure water? Must the labeled water be considered as a solute? How can diffusion occur in a system which is already in thermodynamic equilibrium in the classical sense? The first generalization, of course, is consideration of the process of "self-diffusion." This is another special instance of interfusion by which the movement of water molecules among themselves in an equilibrium system could be studied by the use of labeled water.

More important, however, is the description of steady-state systems which are out of thermodynamic equilibrium. We may consider the physical problem, for example, of a vessel of pure water in which thermal convective movement occurs. Of even more interest, perhaps, are the biological systems which accumulate solutes against concentration gradients. In such systems, if labeled S is released in a region of low concentration and interfusion is allowed to progress, the forces maintaining the high concentration of S in cells or tissues will act on labeled and unlabeled S in the same way (we are neglecting possible isotopic differences). Labeled S will eventually move "uphill" against the concentration gradients in the system. Can this situation be considered as diffusion in its current sense?

In the process of interfusion, the current of labeled material is linearly related to the gradient of the specific activity rather than to the gradient of the concentration. Introducing an "interfusion constant" $\mathscr{I}$ the relation is expressed in the following equation:

$$\mathbf{J}^* = -\mathscr{I}\,\nabla a = -\mathscr{I}\left(\frac{\partial a}{\partial x}\,\mathbf{i} + \frac{\partial a}{\partial y}\,\mathbf{j} + \frac{\partial a}{\partial z}\,\mathbf{k}\right) \tag{1}$$

where ∇ is the gradient operator and $\mathbf{i}$, $\mathbf{j}$, and $\mathbf{k}$ are the unit vectors parallel to the x, y, and z axes. An expression of this type was derived by Sheppard and Householder (4). Figures 48 and 49 indicate schematically the difference between the diffusion and interfusion concepts.

Generalization of the Interfusion Equation

The interfusion eq. 1 is no more a general equation than the equation for Fick's law. It is possible, however, to consider how it may be generalized in a number of ways. As one example, if the system is anisotropic, it is possible to consider $\mathscr{I}$ as a dyadic just as it is possible to generalize the diffusion process to describe diffusion in anisotropic media. In this way, we might use the relation to describe self-diffusion in crystals where diffusion rates may be different in different directions. A second generalization is to remove the restricting assumption of the steady state. Now S may be moving with some particular velocity at P. Consider at some instant, the situation at P from two possible vantage points. If, in one case, we move along with S, we will observe the effect of the movement of label as though it were mingling simply by interfusion. The expression which then applies is eq. 1.

If, on the other hand, interfusion is prevented for the moment and observation is made from a stationary coordinate system, relative to the actual system in which S is now moving, the flux of label will be the flux $\mathbf{J}$ of total S multiplied by the specific activity a, representing the fraction of S which is labeled. If both motions are permitted, we then have for the total flux the sum of the two fluxes, namely

$$\mathbf{J}^* = -\mathscr{I}\,\nabla a + a\mathbf{J} \tag{2}$$

The sum must, of course, be determined vectorially.

From the preceding discussion it may be evident that the concept of a single isotropic constant diffusion coefficient is a gross over-simplification. Furthermore, even in simple diffusion of gases at constant temperature, the resulting data do not provide the most direct approach to the estimation of basic molecular properties. In liquid systems the effect of drag forces between different solutes or between solute and solvent may require consideration. The rate at which labeled water penetrates a porous

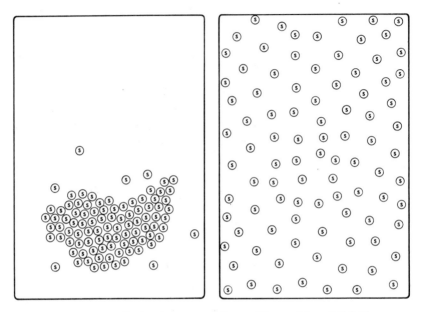

Fig. 48. Schematized diffusion in a compartment: The particles of S (left) move at random until their mean concentration is uniform throughout the container (right).

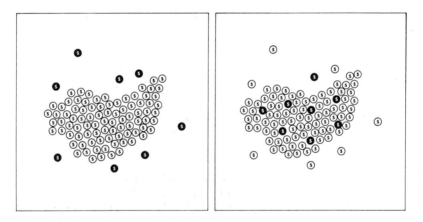

Fig. 49. An example of schematized interfusion. Particles of S are constrained against free diffusion. Particles of labeled S (black) intermingle with unlabeled (white) until the specific activity on the average approaches uniformity (right).

membrane can sometimes vary greatly with the rate at which the total flow of water proceeds through the pores (59). In the analogous case of interfusion the extent to which estimates of $\mathscr{I}$ from tracer data may yield useful information must await considerable further work. It would seem at this moment that this approach to tracer methodology does not promise rapid dividends. The value of eq. 2 appears primarily to lie in its descriptive use. Here we already have a worthy precedent in the use of the Fick law equation of diffusion.

Heat-Flow Analogy

The literature of physics contains considerable material devoted to the study of heat conduction. A large number of physical systems satisfy the requirements that they are isotropic and that their physical properties such as density and thermal conductivity, which are constant with respect to time, vary only slowly with spatial coordinates and with temperature. Within and at the boundaries of such systems the temperature follows the well-known Fourier equation,†

$$\frac{\partial T}{\partial t} = k\,\nabla^2 T \tag{3}$$

By matching conditions at boundaries the analysis can be extended to several such systems in thermal contact. If we accept the Fick-law analogy (eq. 1), a similar equation applies to steady-state tracer systems. Let us consider the previously derived steady-state interfusion equation,

$$\mathbf{J^*} = -\mathscr{I}\,\nabla\mathbf{a}$$

Consider an elemental volume dv and evaluate the surface integral of $\mathbf{J^*}$ over its surface. By a well-known theorem of vector analysis (60) this integral is equal to the integral of the divergence of the vector throughout the volume element. The surface integral represents the net rate at which label is leaving the element. This in turn equals the negative rate of change of label in dv or $-\{\partial([c]a)/\partial t\}\,dv$. Where the concentration $[c]$ is constant in a steady-state system we take the divergence on the right and divide both sides by $[c]\,dv$ obtaining the relation:

$$\frac{\partial a}{\partial t} = \left(\frac{1}{[c]}\right)\text{div}\,(\mathscr{I}\,\nabla a)$$

$$= \left(\frac{\mathscr{I}}{[c]}\right)\left[\frac{\partial^2 a}{\partial x^2} + \frac{\partial^2 a}{\partial y^2} + \frac{\partial^2 a}{\partial z^2}\right] = \left(\frac{\mathscr{I}}{[c]}\right)\nabla^2 a \tag{4}$$

† Here, on page 158, and in Figs. 50 and 51, T represents temperature. For the normal use of T to represent time delay, see the Table of Selected Symbols and Abbreviations. The use of T for time in the previous chapter is dictated by the requirements of the FOR TRANSIT system (see Table 4).

This equation requires that the variation in $\mathscr{I}$ with space coordinates be small enough to permit neglecting the partial derivatives of $\mathscr{I}$ with respect to x, y, and z. Also, the time variation of $[c]$ must be negligible compared to that of a.

The most common application of this equation to heat-flow problems is with respect to systems whose physical properties are uniform throughout. In this instance the mathematical methodology is well developed, enabling solutions for a wide variety of cases to be found in the literature (61). Although thermal analogs for tracer experiments may be of limited use at present, there is a wide literature on the study of heat flow by electrical analogs which may be of value. A recent study by Paschkis and Hlinka provides a representative example (62). Because of the parallelism in the mathematical formulation of heat flow and of diffusion problems the literature on the two subjects has tended to proceed hand in hand.

The field of tracer theory was also brought into the same context by Harris and Burn (63) whose equations were similar in form to those which occur in heat-flow problems. These workers made a theoretical analysis of tracer movements in isolated frog muscle in which both the rate of crossing cell boundaries and of diffusion in the extracellular space were considered.

Lumping in the Solution of Heat-Flow Problems: Application to Tracer Systems

The use of resistance-capacitance elements in electric analog computation is based on an essentially lumped concept and for this reason when analog computers are used in heat-flow problems the solution is frequently approximated by converting the continuous system to a lumped one. As an illustration of how this is achieved, we consider the two-dimensional heat-flow problem in Fig. 50. Here a semi-infinite block of material of low thermal conductivity raised to an initial temperature $T_1(0)$ is placed in contact with an identical block at a temperature $T_2(0)$. The temperature distribution at times t_1, t_2, etc., following contact is indicated schematically in the top half of the figure. The lumped approximation is indicated in the bottom half and a typical electrical analog circuit at the bottom of Fig. 51.

Of course it is conceivable that some type of leaky dielectric system might be constructed in order to obtain a continuous distribution of capacitance and resistance for analog purposes. Usually, however, a sufficiently good lumped approximation is achieved by using capacitances

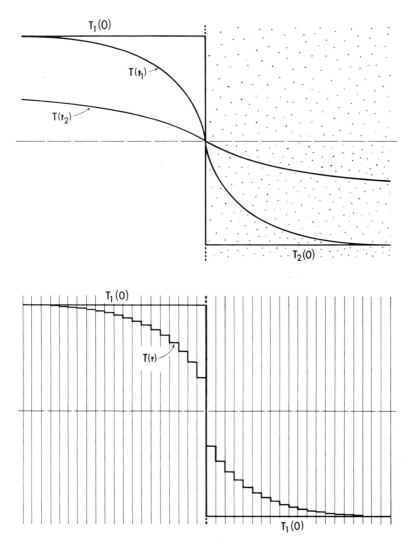

Fig. 50. Continuous heat flow (top). The curves represent schematic temperature distributions at different times. Below is shown the equivalent lumped approximation of the system.

connected by resistors. The equivalent lumped thermal problem concentrates the resistance to heat flow in the insulating laminae between lumped elements whose conductivity and thermal capacity are large.

Figure 51 shows at the top how, by progressively decreasing the lumping interval near the boundary a closer approximation to the true situation is

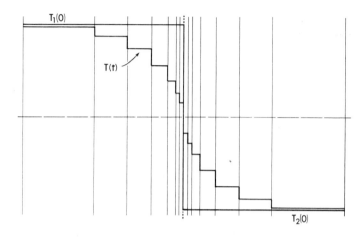

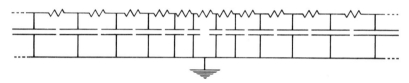

Fig. 51. Effect of varying the lumping interval (top) and a suggested lumped electrical analog (bottom).

achieved for a given number of elements. The bottom of the figure indicates how the equivalent electrical lumping is similarly adjusted.

Integral Equation Description of Tracer Theory: The Age Concept

An equation which includes among its terms an unknown function under an integral sign is known as an integral equation. Thus, for example, the equation

$$f(x) = \int_a^b K(x, t)f(t)\, dt \tag{5}$$

is an integral equation. For certain values of a and b and certain types of functions K, the task in obtaining a solution is to determine a function $f(x)$, if any, which satisfies the equation. The basic theory of linear differential equations is in a well developed state, whereas the study of integral equations represents a more recent branch of mathematical science. For this reason whenever a problem in tracer theory can be adequately treated by a differential equation formulation, this approach has been adopted in the present volume. Nevertheless, there are a few instances where integral descriptions leading to integral equations have suggestive applications. Such a formulation of tracer theory in metabolizing systems was adopted from a partially intuitive point of view by Branson (64, 65). Later criticisms appeared, including a discussion of the mathematical limitations of the theory (66, 67). Recently a good discussion of the mathematical basis of the integral equation approach was given by Stephenson (68).

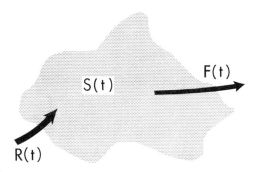

Fig. 52. A general closed tracer system showing the amount of S in the system as $S(t)$, the loss $F(t)$ and the renewal $R(t)$.

As originally formulated, the theory applied to the amount $M(t)$ of a metabolite contained within a system taken as a function of time. The rate at which it entered the system was $R(t)$ and its disappearance from the system was expressed in terms of $F(t)$† which was termed the "metabolizing function." Tracer theory was considered in discussing the role of tracers in determining the form of the functions which described the kinetic processes in the system.

Consider a general closed system (Fig. 52) in which material can first enter in any way by crossing the boundary and then leave by any similar general process. This process may include not only the crossing of the boundary but also the possible conversion into some substance considered no longer to be a part of the system. Suppose that an amount of material S enters the system at zero time. We may define a function $F(t)$ to be the fraction of this initial S which remains at time t. Then, if $S(0)$ is the amount initially placed in the system, $S(0) \times F(t)$ will be the amount

---

† Note here $F(t)$ represents the fraction remaining within the system. This is different to $F(t)\, dt$ elsewhere in this volume which represents the fraction appearing at the outflow in the instant from t to $t + dt$.

remaining at time t. These ideas can be extended to include material entering at times other than $t = 0$, but in order to make this extension it is necessary to clearly understand the situation.

In effect, we are postulating that the material present in the system at any time is the sum of the residua from an independent set of parcels of S. These parcels are delivered into the system at various times, and each behaves as though the other parcels have not been introduced. In this regard we must consider $\mathbf{F}(t)$ as an expression of the "age" of the parcel. If the parcel is introduced at $t = 0$, the age is the same as the time t. However, if introduction occurs at $t = t'$, then the age is $t - t'$. We note, of course, that all ages are positive or that $t > t'$. The fate of each parcel will depend only on its age. Therefore, molecules will not have the same probable fate unless they either belong to the same parcel or to parcels of the same age.

Under these restrictions we may express the amount of S present in the system at any time t in terms of the initial amount* and the integral of the contribution of S from times t' taken over all values of t' up to the observation time t. Thus

$$S(t) = S(0)\mathbf{F}(t) + \int_0^t R(t')\mathbf{F}(t - t')\, dt' \tag{6}$$

where $R(t')$ is the rate at which S is entering the system at the moment when $t = t'$. The convolution integral was already introduced on pages 85–97 of Chapter 5. It is more extensively discussed on pages 204–206 of Chapter 9. Although a wide latitude is permitted for $R(t)$, difficulty may at times be encountered with the integral equation approach if $\mathbf{F}(t)$ is not a stable function of t. In that event what one parcel of a given age does in the system may not be the same as what another parcel of the same age does. In "stationary" systems however, as a result of constant properties, all parcels of the same age should behave alike irrespective of the time of introduction into the system. One such example is the flow of blood through a stable circulatory labyrinth. Here, injected labeled erythrocytes may have the same historical behavior regardless of the moment of injection. Information about such systems may be obtained by injecting a single labeled parcel and then observing the variation in amount of emerging tracer as time progresses. A "non-compartmental" analysis which is somewhat similar in form to Branson's has been given for thyroidal radioiodine by Nadler (69).

---

* As Hearon and Wijsman indicate, all S in this "initial amount" must be of the same age. In metabolizing systems it is difficult to understand why such an amount should suddenly be introduced at $t = 0$. In the case of injection of tracer into the circulation, however, there may be some value in retaining $S(0)$ in the analysis.

It was pointed out by Stephenson (68) that we may define a "turnover time" and a "pool size" in such a non-compartmental system. The "turnover time" is merely the arithmetic mean time which a particle of S spends in the system, namely

$$\bar{t} = \int_a^b tF(t) \, dt \tag{7}$$

In this relation $F(t) = d\mathbf{F}(t)/dt$, is the fraction of label emerging from the system between t and $t + dt$. If R is constant, then the "pool size" is simply the product of the rate at which S flows through the system times the average time the particles spend in it. Thus

$$S_{\text{system}} = R\bar{t} \tag{8}$$

This approach to tracer theory will receive further discussion in Chapter 9 in connection with studies in the circulation.

Sources and Sinks in Tracer Equations

In general, a closed system would contain no sources and sinks of S or of tracer. However, in some instances, simplification may be achieved by isolating one part of a closed system and considering it to interact with the remainder. Labeled S might then enter from the surroundings through one or more sources and leave through one or more sinks. Stephenson (35) has described this situation for solutes using the classical transport equation

$$\nabla \mathbf{J_k} = -\partial[c_k]/\partial t + s_k \tag{9}$$

where $\mathbf{J_k}$ is the flux of the kth species, $[c_k]$ is the concentration and s_k is the "source strength" i.e., the net production per unit volume. These quantities are functions of the space and time coordinates. In the case of the usual discrete sources s_k will be positive at certain points of space and zero elsewhere. For sinks s_k will be negative.

In terms of specific activity and labeled S if the concentration $[c]$ of S is constant we have

$$\nabla \mathbf{J}^* = -[c]\partial a/\partial t + as \tag{10}$$

where s is the total source strength of S and a is the specific activity of entering labeled S. In the case where s is zero this expression may be reduced as before to eq. 4. As shown by Stephenson equations of this type may be integrated to obtain compartmental expressions, using standard theorems of vector analysis.

Membrane Processes

The importance of the problem of membrane transport in biology (70) has recently led Nims (71) to review it from the standpoint of fundamental physical chemistry in an elegant manner. The problem is instructive from the tracer theoretical point of view since it provides a point of contact between the theories of continuous and of compartmental systems. Those who wish to pursue the subject further into its basic biophysical implications will consult the relevant literature. We will, for simplicity, consider only membranes separating stirred compartments. We will also confine our attention to the post-transient state and will postulate that there are no sources and sinks in the membrane.

From eq. 2 we have

$$\mathbf{J}^* = a\mathbf{J} - \mathscr{I} \, \nabla a$$

In this system $\mathbf{J}$ will be everywhere zero in the body of each compartment since random movements due to the stirring will average out net movements. Also the stirring will cause ∇a to be zero. We choose the x axis perpendicular to the membrane whose thickness will be Δx. We also assume there are no components of $\mathbf{J}$ or gradients of a parallel to the membrane. Thus, throughout the membrane using x components J_x and J_x^*,

$$J_x^* = aJ_x - \mathscr{I} \frac{\partial a}{\partial x} \tag{11}$$

If we replace $\partial a / \partial x$ with $\Delta a / \Delta x$ and $\mathscr{I}$ with $M_e RT \, \Delta X$ of the Nims theory, this relation has the same form as the second equation 36 in Nims' paper. Thus, Δx in the Nims theory corresponds to the mean distance over which molecules exchange in the present theory. $M_e RT$ corresponds to the mean rate of exchange. In more general situations the relation between $\mathscr{I}$ and fundamental physical quantities may be less obvious. (In the Nims theory M_e is the "exchange coefficient," R is the gas constant and T the absolute temperature).

We will consider only the situation where initial transient fluctuations in concentration and specific activity have subsided. At this ultimate time, since after a certain period of time neither S nor tracer are accumulating or being lost in the membrane, both J_x^* and J_x must be constant. If there is a difference in specific activity between the two compartments the variation of a with x must be balanced by a corresponding variation in $\partial a / \partial x$.

The differential equation which results (since after the transients have disappeared a is not a function of t) is

$$-\mathscr{I}\frac{da}{dx} = J_x{}^* - aJ_x \tag{12}$$

which has for a solution

$$a(x) = \left(a_1 - \frac{J_x{}^*}{J_x}\right)\exp\left(\frac{xJ_x}{\mathscr{I}}\right) + \frac{J_x{}^*}{J_x} \tag{13}$$

where a_1 is the specific activity in compartment 1 and x is measured from the interface between the membrane and compartment 1. By inserting a_2 for $a(x)$ and Δx for the membrane thickness, we invoke the requirement that $a(x) = a_2$ in the second compartment. It is now possible to solve for $J_x{}^*$ obtaining

$$\frac{J_x{}^*}{J_x} = \frac{a_2 - a_1\exp\left(\Delta x\frac{J_x}{\mathscr{I}}\right)}{1 - \exp\left(\Delta x\frac{J_x}{\mathscr{I}}\right)} \tag{14}$$

These relations would of course only be approximate in most actual thin membranes since the variation of $\mathscr{I}$ with space coordinates would be subject to variations in the architecture of the membrane which might cause $\mathscr{I}$ to be discontinuous in some cases. As pointed out by Sheppard and Householder the validity of eq. 4 and other similarly derived relations requires that a not vary rapidly over distances of the order of mean distances between interactions of molecules. It is clear, however, that there should not be a discontinuous jump in a across a membrane. If $a(x, y, z, t)$ does not measure the actual specific activity at a mathematical point it may to a better approximation provide a measure of the probability that a molecule of S at the point be a labeled one. Of course this illustrative analysis only predicts $a(x)$ if a_1, a_2, Δx, J_x, and $\mathscr{I}$ are given. The theoretical and experimental problems relative to estimating basic properties of membranes from tracer kinetic data and other information are beyond the scope of this volume.

In his review Nims makes the statement: "a tagged component 'traces' the flux of a normal component when, and only when, the molar ratio of the two species is the same in all compartments." Lest this statement create misunderstanding, his definition of "flux of a normal component" represents only the first term on the right of eq. 11. This term is the bulk transport and does not include transport by exchange. The bulk component is the difference between the rate from 1 to 2 and the opposing rate. If *both rates* are determined by the tracer method then the bulk component may be obtained by difference.

9

Tracers and the Study of the Circulation: Stochastic Processes

The tracer method has been familiar to the physiologist in studies of blood circulation and body fluids for more than half a century. Since the development of isotopic tracers, improved experiments have been conducted which include the labeling of erythrocytes with radioactive iron (72), and with isotopes of phosphorus chromium and selenium. Nevertheless, much work has been done and continues to be done using organic dye labels with which plasma can be traced and blood flow studied. In physiological research, then, it is particularly important to regard the tracer method from a general viewpoint. Because the literature on the purely experimental aspects of tracer research in the circulation is prodigious, the discussion will be confined to a rather brief presentation supplemented by references to other more detailed reviews (73) for a more comprehensive introduction to the subject.

Laminar Flow in Long Uniform Tubes

As an introduction we may consider the transport of label by a Newtonian fluid moving at low Reynolds number through a long cylindrical tube of circular cross section. Here, if a small uniform bolus of labeled material is introduced and followed, the situation is a completely deterministic one. Flow is "parabolic" with fluid stationary at the wall and flow maximal in the center. The problem is illustrated in Fig. 53, and the coordinates used in the derivation are also shown. An analysis will be outlined for a very short bolus from which, by integration other situations can be discussed.

More lengthy and complete analyses have been given by several workers (74, 75, 76, 77).

In deriving the theoretical relation we will adopt the following notation— some of which can be found in the Table of Selected Symbols and Abbreviations:

$r, v(r)$ = radius and velocity respectively of a given laminar sleeve of fluid. The velocity is considered to vary with r.

R, L, V = radius, length, and total volume of the tube respectively.

Q = volumetric flow rate through the tube.

$\tau = Qt/V$ = fraction of system volume displaced in time t.

I = amount of label in the initial bolus.

$C(t)$ = concentration of label in samples collected at the outflow between time t and $t + dt$.

$F(\tau)\, d\tau$ = fraction of label collected at the outflow between τ and $\tau + d\tau$.

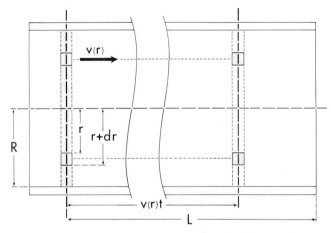

Fig. 53. Bolus movement in a cylindrical tube showing coordinates.

We may resolve the fluid in the tube into a series of laminar sleeves. For an initially short bolus the label may be resolved into a corresponding series of laminar rings (Fig. 54) of variable radius r and thickness dr with variable velocities $v(r)$. Consider a particular ring arriving at the outflow at time t. Since the bolus is uniform in length and concentration, the amount of label per unit cross-sectional area entering each ring will be constant everywhere. To relate the amount of label which this ring delivers at the outflow to the time of delivery, we recall the following basic relation of parabolic flow for the fluid sleeves:

$$v(r) = \frac{2Q}{\pi R^2}\left(1 - \frac{r^2}{R^2}\right) = \frac{L}{t} \tag{1}$$

By differentiation we obtain

$$-dv = \frac{L \, dt}{t^2} = 4Qr \frac{dr}{\pi R^4} \tag{2}$$

The fraction of total label contained in the ring is the ratio of its cross-sectional area (taken as an absolute value) to the cross-sectional area of the total tube. Thus the amount of label in the ring is

$$\frac{2Ir \, dr}{R^2} = \pi R^2 L I \frac{dt}{2Qt^2} = IV \frac{dt}{2Qt^2} \tag{3}$$

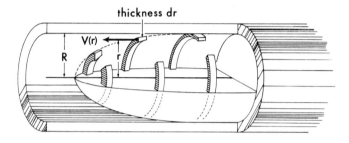

Fig. 54. Dispersal of a bolus into infinitesimal rings by parabolic laminar flow.

On dividing the amount of label emerging as the ring reaches the outflow by the volume of fluid flowing out in the infinitesimal interval, namely $Q \, dt$, we obtain for the concentration

$$C(t) = IV/2Q^2t^2 \tag{4}$$

However, we note that the velocity is maximal at $r = 0$; therefore $v_{\max} = 2Q/\pi R^2$. Correspondingly the time for earliest appearance of label is $L/v_{\max} = V/2Q$, therefore $C(t) = 0$ for all times earlier. Since the interval between τ and $\tau + d\tau$ is equivalent to the interval between t and $t + dt$, the fraction of injected label which emerges between t and $t + dt$ is

$$\frac{C(t)Q \, dt}{I} = \left(\frac{V^2}{2Q^2t^2}\right)\left(\frac{Q \, dt}{V}\right) = F(\tau) \, d\tau \tag{5}$$

Thus for a very short bolus

$$\begin{aligned}
F(\tau) &= 1/2\tau^2 &&\text{for } \tau \geq 1/2 \\
&= 0 &&\text{for } \tau < 1/2
\end{aligned} \tag{6}$$

In verifying this relation experimentally, we must balance the labeled material carefully against the unlabeled so that gravitational separation of the two fractions will not occur. Simple injection of the bolus is not sufficient to produce a satisfactory initial parcel of label. When the system is controlled properly, however, a fairly good agreement can be obtained experimentally with the theoretical relation (Fig. 55). We must note that the theoretical relation is derived on the basis that it is the mean flux of label which is observed at the outflow. This condition is the case when emerging fluid is recovered in a series of small sampling tubes and the label content in these containers is determined.

The previous analysis applies to the situation where the initial distribution of label is obtained by depositing it uniformly spacewise along the tube. If, however, laminar flow occurs in the region of the tube entrance

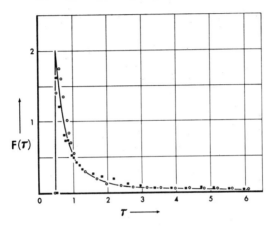

Fig. 55. Comparison of theory (solid line) and experimental points obtained at 0.37 cc per sec (circles) and 0.85 cc per sec (squares). (Data from Sheppard et al. (76); reproduced by permission of the American Heart Association.)

(a situation roughly true if a long thin tube is connected to a bed of glass beads), then a situation occurs where label may be "turned on and off" very briefly. Thus the distribution is uniform at the inflow when taken with respect to time coordinates rather than to space coordinates. In this case central laminae will receive more label than will peripheral ones. Therefore, since the more rapidly moving laminae contain label in proportion to the velocity, the previous expression for the fraction of label in a given sleeve must be multiplied by the velocity of the sleeve and divided by the average velocity as a normalizing factor. The amount of total label contributed in any interval dt is now

$$\frac{2Ivr\,dr}{R^2v_{av}} = \frac{\pi R^2IVv\,dt}{2Q^2t^2} = \frac{IV^2\,dt}{2Q^2t^3} \tag{7}$$

where v_{av} is the mean velocity $= Q/\pi R^2$. In terms of τ variables

$$F(\tau) = 1/2\tau^3 \qquad \text{for } \tau \geq 1/2$$
$$= 0 \qquad \text{for } \tau < 1/2 \tag{8}$$

It is seen that ideally one obtains a $1/2\tau^2$ or a $1/2\tau^3$ law (Figs. 56 and 57), depending upon the weight, i.e., relative observational importance, assigned by the physical situation to the various sleeves in proportion to their flow velocity. Now, instead of mean flux at the outflow, it is possible under certain circumstances to observe mean concentration in the tube. For example, a tube may have a thin wall and be wrapped around a G.M. counter in such a way that the emission of highly penetrating

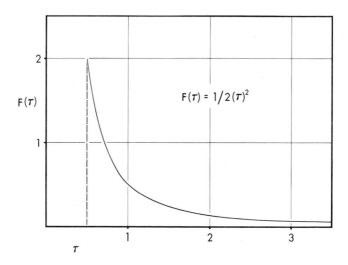

Fig. 56. The $1/2\tau^2$ law of bolus washout.

radiations will affect the detector equally, independent of the origin of the radiation within the tube. Here all laminae contribute equally to the effect on the detector whether they move rapidly or slowly. We must, in comparison to the previous situation, apply the weighting procedure in reverse so that a $1/2\tau^3$ relation may be converted to a $1/2\tau^2$. These considerations have been discussed more completely by Sheppard et al.* (76) and by Rossi et al. (77). Of course, in many practical cases the negative power of τ which applies may not be an exact integer, thus requiring some empirical fractional power in order to fit the data adequately.

A further interesting concept arises if diffusion processes occur between laminae. This situation may occur with highly diffusible substances such as small colored inorganic or organic molecules in small tubes at low flow

---

* The author is indebted to Dr. José González-Fernández for recently pointing out that a turbulent mixing device should not convert the outflow from a glass bead column and serially connected tube from a $1/2\tau^3$ relation to a $1/2\tau^2$ relation.

rates. Taylor (74) has given a comprehensive analysis of this type of effect. In general, the sharp upsweep and the almost discontinuous change in the derivative at the peak become smeared out, yielding from the shape of the curve an estimate of the value of the diffusion constant of the indicator substance. In studies with labeled plasma, the dyes commonly employed adhere tightly to the plasma protein molecules and these diffuse slowly.

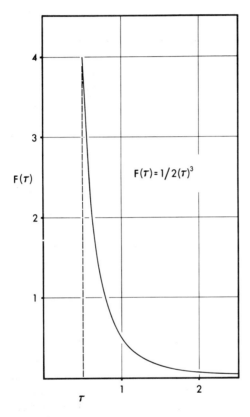

Fig. 57. The $1/2\tau^2$ law of bolus washout.

Nevertheless, in the presence of erythrocytes (red blood cells) which have a strong jostling movement in flowing blood, a situation may conceivably exist in which plasma might undergo a pseudo-diffusion effect as a result of microturbulence produced by the erythrocyte motions. Consideration must also be given to the appreciable deviation from Newtonian flow which may be observed in small tubes perfused with blood containing a high concentration of erythrocytes.

Non-Deterministic Processes

In the previous section we were concerned with the fraction of injected label appearing in a given time (or τ) interval at the outflow when a small highly concentrated bolus of label was injected at the inflow. In this case the distribution in traversal times was the result of completely deterministic processes within the system. In many instances, however, attempts to follow in detail the movement of every element of dye or other label may be fruitless. It will then be necessary to consider the system as a chaotic one in which the processes are random in nature. This is the philosophy employed in statistical physics in which the mechanical behavior of single gas molecules is ignored and analysis is confined to the overall statistical properties of the population. In this method, then, it is necessary to deal with statistical distribution and density functions. Just as we may start a group of runners at some instant and consider their different arrival times at the end of the race course so we must now consider mathematically the distribution in transit times of label through the system. The analysis must begin with a consideration of the mathematical description of "simultaneous" start of labeled material.

The "unit spike" and the delta function.   In the previous section we introduced the concept of a bolus of labeled material and in Chapter 6 we discussed the electrical analog of this as a high voltage "spike." In general, we may wish to consider the mathematical expression of a physical quantity which has a very high magnitude for a brief interval of time. Of course, in terms of physics, what we are actually concerned with is a voltage or a dye concentration which is of finite (though possibly very large) magnitude and not vanishingly small in width. For this reason we will draw it in our illustrations in this volume as having a certain amount of width at the base. This will be the imperfect physical realization of a mathematically ideal concept which we define as a "delta function." This peculiar mathematical function is basically discontinuous. It has an infinite ordinate value for a single mathematically discrete abscissal value. Usually, its integral is considered to represent a unit area. If the abscissa value is a where the ordinate departs from zero the function is commonly represented as $\delta(t - a)$. Often the function will be used to represent an impulse at $t = 0$, in which case we write $\delta(t)$.

Consider any distribution of concentration taken as a function of time. Let the area under the curve be unity. Now imagine that the horizontal scale is contracted and the vertical scale increased (Fig. 58). It is possible under a fairly broad set of mathematical conditions to imagine that the interval embracing all values of the independent variable, which might be t, for example, can become infinitesimally small and the ordinates infinitely large in such a way that the unit area condition is maintained. The genesis of a delta function and the concept of unit area under it must thus be considered in terms of a mathematical limiting process. It is of theoretical interest that a variety of peaked functions may be regarded as passing to a

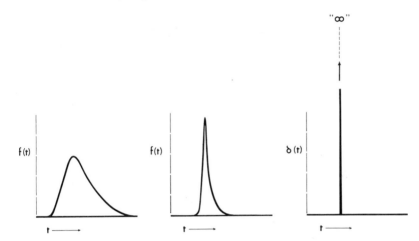

Fig. 58. Schematic illustration of the genesis of a "delta function."

delta function in the limit. In other words any physically reasonable label distribution which is of high enough concentration and short enough in duration may be thus approximated.

In terms of physics, if all dispersion functions $f(t)$, etc., are broad when compared to the width of the spike, then an imperfect spike of finite height and width will be an adequate approximation of the situation. Such a spike which has unit area under it will be termed a "unit spike." It will, of course, be necessary only to know the response of a system to a unit spike since other responses are obtained by multiplying by the area under the stimulus curve. In the case of tracer experiments this area is the integral under the concentration-time curve of the injectate—namely, $C_0(t)\Delta t$. Normally it is the area of the injectate spike rather than the concentration which is of paramount interest. If the response to a spike is known it is then possible to obtain the response to other stimuli by integration.

It is a problem for the advanced mathematician to determine under what conditions it is possible to resolve a continuous distribution function into a volley of elementary spikes for the purpose of integration. It is sufficient for the present to indicate how this task is done in typical problems of a physical nature (Fig. 59). Consider the ordinate of a primary curve such as that of the concentration of label entering a circulatory labyrinth. Imagine a rectangle whose height is the ordinate $C(t)$ and whose width is the infinitesimal dt. Of course, this elementary area is vanishingly small, and there is an infinite number of them. The curve area is finite and

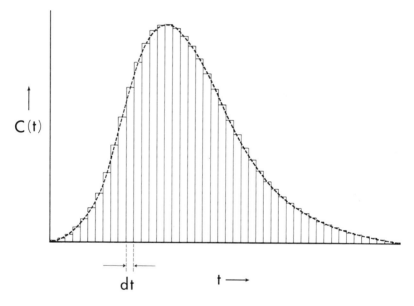

Fig. 59. Schematic illustration of the resolution of a dye curve into a series of elementary pulses. Ordinate is concentration, abscissa is time.

after $C(t)$ has been modified by its passage through the labyrinth, the area remains finite. Let us regard the entering label as a succession or volley of elementary impulses $C(t)\,dt$. If we are mathematically able to describe the response of the labyrinth to one of these impulses (delta function response), then, because of the additive property of label at the outflow, we merely sum all of the elementary responses to obtain the overall response to the succession of impulses. It is the response to a delta function stimulus, then, which we may seek if we wish to characterize the system from the basic point of view.

The mathematical significance of the delta function may be further appreciated by considering its relation to the unit-step function which is also commonly encountered in physical problems. The unit step is also a discontinuous function which is zero everywhere for all values of the independent variable less than a and has a constant value of unity for values equal to a or greater. Practical illustrations of this relation were given in Chapter 6 (Fig. 27A). The function $\delta(t - a)$ may be obtained as the derivative of the unit step. Just below and above the edge of the step at a the derivative is zero, but at a the derivative suddenly becomes infinite. Of course, in terms of physics, the unit step is an idealization since some rounding of the corners of the step will always be detected in any physical situation provided sufficiently precise observations are made. Further discussion of these functions may be found in more advanced texts (39).

Bolus response versus infusion response. We have noted above that a function representing a physical variable may be represented as a sum or integral of delta functions. Thus if $F(t)$ is the response of a linear system to a delta function the response to a step at $t = a$ may be represented as

$$\int_{-\infty}^{t} F(t - x)\, dx = \int_{a}^{t} F(t - x)\, dx = \int_{0}^{t-a} F(t)\, dt \qquad (9)$$

since $F(t - x) = 0$ for all values of $x < a$. Practically speaking in tracer theory the response $F(t)$ may be considered as the response at the outflow to a bolus injection of label and $\int_{0}^{t-a} F(t)\, dt$ is the response to an infusion of label starting at $t = a$. Since one response may be derived from the other there would seem to be little theoretical choice between them. In the language of the statistician $F(t)$ represents a "frequency density" function and $\int_{0}^{t} F(t)\, dt$ represents a "frequency distribution" function. This will be the response to a step at $a = 0$.

An important relation whose proof may be found in more advanced treatises may be illustrated in the case of the delta and step functions. We define the Laplace transform of $\delta(t)$ as

$$\mathscr{L}\delta(t) = \int_{0}^{\infty} e^{-st}\delta(t)\, dt = e^{-0t} = 1 \qquad (10)$$

since $\delta(t)$ has a value only at $t = 0$.

Similarly if $U(0)$ is a unit step at $t = 0$

$$\mathscr{L}U(0) = \int_{0}^{\infty} e^{-st}U(0)\, dt = \frac{e^{-st}}{s}\Big|_{0}^{\infty} = \frac{1}{s} \qquad (11)$$

In general the transform representing a differential response may be converted to the transform for an integral response from $t \doteq 0$ by the factor $1/s$, provided both transforms exist. From this viewpoint in circulation studies bolus or infusion experiments differ by this factor in transform space.

The Stewart-Hamilton Principle

Blood flow. If the concentration of label in a fluid passing some observation point is constant over the cross section of the vessel, then a condition of uniform mixing exists such that any sample of the fluid is a valid sample of all of it passing the point. Consider the concentration $C(t)$ in a small sample taken in an interval of time Δt so small that the concentration does not appreciably vary. Then, if the total amount ΔI of label in the sample is known it is possible by the principle of simple dilution to know the volume ΔV of the sample. From the volume ΔV and the collection time, the flow rate Q can be computed as

$$Q = \frac{\Delta V}{\Delta t} = \frac{\Delta I}{C(t)\,\Delta t} \tag{12}$$

Of course this relation is truly precise only if the limit is taken over a time interval that is infinitesimal. Finite, though small, intervals and equivalent mean values of $C(t)$ can be used in practice. Suppose that label is delivered to the fluid at a known uniform rate, and that the fluid flow is uniform. After the system has reached equilibrium the plateau concentration of label in the fluid obviously yields the flow rate by the same type of dilution principle. However, even if the concentration is not uniform with time, it is still possible to determine the rate, provided there is some means at hand to determine ΔI.

As an illustration consider the situation (Fig. 60) where $C(t)$ can be determined in blood flowing out of the aorta following injection of a small bolus of highly concentrated label into the right heart. Here pioneer studies by Stewart were developed into a valuable physiological tool by Hamilton and his coworkers (78). If there is no recirculation and the area under the entire curve of $C(t)$ can be determined, then the label in any small interval Δt is to the total label as the area of the curve from t to $t + \Delta t$ is to the total area. Thus, if I is the total label introduced, the amount in the small interval is $\overline{IC(t)}\,\Delta t \Big/ \int_0^\infty C(t)\,dt$, where $\overline{C(t)}$ is the average concentration in the small interval. Dividing the amount of label in the interval by the average concentration yields the volume of fluid included in the interval. Division of this volume by the time duration Δt yields the flow rate. Thus

$$Q = \frac{I}{\int_0^\infty C(t)\,dt} \tag{13}$$

A review of the use of this relation in determining blood flow has been given by Dow (73). The only postulates which are required are that the concentration of the label in the samples truly reflects the concentration in the outflowing blood, that the entire amount of indicator introduced emerges or that the percentage loss in the system is known (permitting the use of a "corrected" I), and that the flow be constant during the interval over which $C(t)$ is described. If Q is not constant, the second law of the

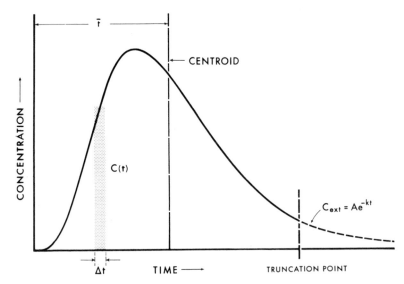

Fig. 60. A schematic dye curve on rectangular coordinates showing dye concentration observed in the aorta as a function of time. An arbitrary differential element of the curve is shown as the stippled bar. The point of truncation indicates how the curve may be prematurely terminated. Note the extrapolation according to an approximate exponential law (dashed portion).

mean may be used to obtain an effective value. The form of $C(t)$ is immaterial, provided that its total integral can be correctly evaluated.

There are practical difficulties in the application of the principle to the determination of blood flow in man. One problem is imposed by recirculation. If the major portion of the first circulation curve can be recognized in the record of label concentration in the aorta or a close tributary up to some truncation point, then Hamilton's ingenious method can be applied to correct for recirculation. Here we may assume as a first approximation that the final tail of the first circulation curve can be approximated beyond the truncation point by a single exponential $C_{ext} = Ae^{-kt}$ (Fig. 60). Plotting the downslope on semilog paper, as far as it can be recognized the curve is linearly continued past the point where recirculation occurs and the recirculated component subtracted. Since in normal man the tail usually represents a small contribution, any error in the assumed linearity is assumed to be of minor importance. Maximum precision in the method will doubtless be achieved when the controversial role of the coronary circulation has been clarified.

The determination of labyrinth volume. It was pointed out many years ago that the product of flow rate and mean transit time of label through the heart and lungs should yield the "central volume" of this system. The mean transit time is obtained from the abscissal value of the centroid of the $C(t)$ curve in Fig. 60. It was reasoned that this is the mean outflow time of the label, i.e., the mean time required for fluid to displace the volume of heart and lungs (and any other volume included in the circuit). Thus, if this time is multiplied by the flow rate Q, the result should yield the central volume. A more general derivation of the principle has been given by Meier and Zierler (79) and Stephenson (68). These derivations have been obtained for the integral form of the curve but, as we have pointed out, this can be simply related to the differential form.

In deriving the principle, we assume that the flow rate Q is known. If the variable t is multiplied by Q, a new variable V results which is the volume of fluid which has emerged from the labyrinth since injection of the label. We may imagine that the initial bolus of label (usually dye or radioactive albumin) can be resolved into a very large number of very small elements. As each element arrives at the outflow at a time t, it has been preceded by a corresponding elemental column of fluid which occupied the space ahead of it along its path (Fig. 61). Each element of dye represents a marker of the terminus of the fluid which preceded it. Many elements will arrive simultaneously at the outflow, and the number of elements arriving at a given instant determines the number of elementary columns of fluid which are contributing dye at the same moment. We thus conceive of the outflow curve as representing a scanned history of the washing out of the labyrinth with fluid. Elementary columns of small volume will contribute their dye first; longer elements, later; and finally the most lengthy ones.

The central volume principle is based upon an important postulate which is more restrictive than the requirements for measuring flow. At the inflow the label must be distributed among the various paths in proportion to their relative flow rates. Let this be the case and consider the interval of time between t and $t + dt$. A certain fraction of the paths through the labyrinth are contributing label at the outflow during this interval. Their local volume is equal to the product of their particular local flow rate dQ

and the time t, since t is the time it takes to displace their volume once. Let k be the constant of proportionality between inflowing rate and label

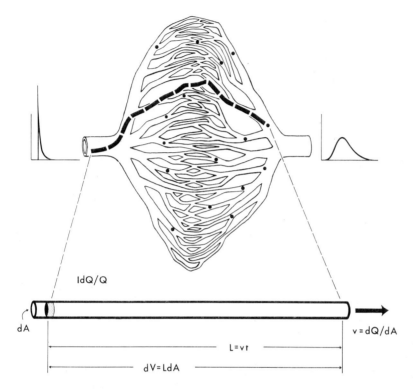

Fig. 61. Schema for deriving the "central volume" relation. Flow is resolved into a population of parallel flow elements of which one is shown. Total label I is partitioned among the elements in proportion to their flow rate, an element whose rate is dQ receiving IdQ/Q units. Arrival of label at time t at the outflow marks the displacement of the element volume $dV = tdQ$ into the effluent.

content. We relate dQ to the label contributed by the local portion of the labyrinth at the outflow at time t; thus

$$dQ = kdI = kC(t)Q \, dt \qquad (14)$$

where the label delivered between t and $t + dt$ is the concentration times the overall outflowing volume delivered in that interval. This outflowing volume is the overall flow rate Q times the interval width dt. Integrating both sides of eq. 14 from 0 to ∞, we obtain $k = 1 \Big/ \displaystyle\int_0^\infty C(t) \, dt$. The volume of the portion of the bed contributing dye (not the outflowing volume) is thus

$$dV = kC(t)Qt\,dt = \frac{C(t)Qt\,dt}{\displaystyle\int_0^\infty C(t)\,dt} \tag{15}$$

The total labyrinth volume is obtained by integration. Thus

$$C.V. = Q\,\frac{\displaystyle\int_0^\infty tC(t)\,dt}{\displaystyle\int_0^\infty C(t)\,dt} = Q\bar{t} \tag{16}$$

where $\bar{t}$ is the arithmetic mean of t taken over the entire outflow curve. It is also, by definition, the abscissal value of the centroid of the curve.

It is valuable at this point to note several concealed postulates in this derivation. Not only is uniform distribution of inflowing label assumed, but it is also assumed that label and fluid move together without slippage. Correction must be made if labeled red cells are used to determine central plasma volume. The reason is that labeled red cells and plasma do not move at quite the same rate through the heart and lungs (80). It is also assumed that label does not trace a given path more than once and that every injected bolus will have the same history as every other. As we will show, on pages 182–184, the concept of a system of stable flow paths, which we have used, need not be invoked.

Basic postulates in circulation studies. It is evident that the results of indicator dilution experiments depend to an important degree on the basic postulates. In determining the flow rate, we usually consider the concentration as being uniform in the stream whose flow is being measured. Many physiological observations have suggested that this postulate is not readily satisfied (81). In the mammalian circulation there is a strong tendency for streamline flow to occur. It is known that streamlines have the notorious property of tending to maintain their integrity over long distances, even in a vascular labyrinth. Label tends to follow these streamlines and not to move between them. Of course, evidence is fairly clear that there is sufficient turbulence in the right heart and aorta to permit the usual method of cardiac output determination to be successfully achieved. In other types of vascular beds, however, considerable difficulty can be expected, and some investigators have preferred to inject their tracer material at high velocity in an attempt to produce turbulent mixing.

When samples are taken through a long cardiac catheter, the dye concentration curve may be distorted considerably. Distortion of the curve, per se, will not produce an important effect on flow measurements, provided that the area is not altered. With catheters of large volume, however, it is important to insure that—if distortion occurs—it does not alter the ratio of the curve area to the amount of label injected. The requirement that $C(t)$ be constant over the outflowing cross section is equivalent to the postulate that the mean flux of label is being measured. When the concentration is not uniform in the outflow cross section any method of averaging it which does not weight the label concentration in proportion to flow velocity will create an error. It is this effect which caused an apparent violation of the Stewart-Hamilton principle in one of the instances discussed by Rossi et al. (77).

When central volume, rather than flow, is being determined, the postulate that the entering dye is distributed among various paths in proportion to flow rate is important. If this postulate is also satisfied, then, even if distortion of the curve occurs, the result will still be valid as long as the centroid of the curve is not affected. Further discussions of the effect of distortion of curves by sampling systems may be found in the literature (75, 76). In general, the effects will be small if the ratio of sampling catheter volume to labyrinth volume or equivalent centroidal volume is small.

The age concept in circulation studies. In Chapter 8 we reviewed the concept of a general closed system into which some substance was moving at a rate $R(t)$ and from which it was being removed at a rate determined by a "metabolizing function" $F(t)$. A vascular labyrinth such as the heart and lungs can be regarded—at least from the tracer viewpoint—as an analogous system. By the use of such a general approach we need no longer restrict the analysis to a system of fixed flow paths. The "pool size" is represented by the central volume and the "turnover time" is equivalent to the mean transit time.

Our approach is based on the analysis of Meier and Zierler (79) with some modifications. Blood enters at a constant rate $R(t) = Q$ and each element moves through the system by processes which can be partly deterministic and partly random. We may imagine the elements as having individual histories. We may introduce the concept of an average history which may be traced by introducing a bolus of label into the inflow and determining the fraction of label appearing in various time or τ intervals at the outflow.

Each element will have an "age" which determines its probability of emerging from the outflow at any given time. The age will depend on the duration of its stay in the labyrinth. We must keep clearly in mind that we are concerned with the distributions of transit times of label only as reflections of the traversal times of the substance being traced, whether it be blood, plasma, or something else of physiological interest.

As fluid enters a labyrinth a certain fraction of it will appear early at the outflow and other fractions will appear later. If the distribution of traversal times does not vary with time, we may consider this as defining a function $F(t)$, and $F(t)\,dt$ gives the fraction of material appearing in a given infinitesimal interval dt at the outflow. This fraction will be proportional to the concentration of an indicator substance at the outflow, provided the indicator adequately serves as a tracer for the fluid being studied. Of course, if different injections of label yield different $F(t)$'s, the basic postulates are violated and we can proceed no further. If the movement through the system is determined by random processes one may regard $F(t)$ as a probability density function which gives the relative chances that the fluid elements will take the various possible times to pass through the system.

We will consider first a system in which the flow rate Q is constant. In this system, as fluid traverses the system, we may consider the fluid elements which have entered after some arbitrary initial time $t = 0$. These new elements will gradually displace the old from the system until eventually all of the system has been displaced. It is, of course, important to consider what is meant by complete displacement. If certain ultimate recesses have not been displaced, these elements of the system will not be included in the volume which is to be measured by the tracer method. If a tracer experiment of the bolus type were being performed, the tail of the curve would then be lost. In a step-function or infusion experiment ultimate terminal concentration would not be achieved at the outflow.

As displacement by new fluid proceeds to completion, we can determine the volume of the system as the difference between the amount of new fluid which has flowed in and that which has flowed out. None of the new fluid elements which have traversal times greater than t will have had time to traverse the system yet. Such elements will remain within the system and will be substituted at the outflow by corresponding elements of old fluid. Since $F(t)\,dt$ is the fraction of traversal times in the interval between t and $t + dt$, the total fraction of times greater than t is

$$\mathbf{F}(t) = \int_t^\infty F(t)\,dt \tag{17}$$

and, the fraction of traversal times less than t is

$$1 - \mathbf{F}(t) = \int_0^t F(t)\,dt$$

This is the fraction of new fluid appearing at the outflow. Since the rate at which new fluid is leaving the system is $Q[1 - \mathbf{F}(t)]$, and the rate of entry is Q the content of new fluid in the system at time t is

$$L(t) = Q \int_0^t \mathbf{F}(t)\, dt \qquad (18)$$

When the total volume is completely displaced we have

$$V = Q \int_0^\infty \mathbf{F}(t)\, dt \qquad (19)$$

This is the system volume. The right hand expression may be simplified by integrating by parts. Thus

$$\int_0^t \mathbf{F}(t)\, dt = t\mathbf{F}(t)\Big|_0^t - \int_0^t t\mathbf{F}'(t)\, dt$$

$$= t\mathbf{F}(t) + \int_0^t t F(t)\, dt \qquad (20)$$

since, by differentiating eq. 17 with respect to the lower limit, we have

$$\mathbf{F}'(t) = -F(t)$$

Proceeding to the limit $\mathbf{F}(t) \rightarrow 0$, thus

$$V = Q \int_0^\infty t F(t)\, dt = Q\bar{t} \qquad (21)$$

There are a number of practical implications to this derivation. For example, the system may contain recirculating paths without violating the principle. This would be true in the cardiovascular system, where the coronary vessels may carry label back into the inflow. The reason for this apparent paradox relates to the fact that if label recirculates before appearing at the outflow the fluid carrying it will also recirculate. This will certainly cause label to appear at the outflow later in time, but the fluid flow associated with it will also be less in proportion so that the associated volume will still be correctly represented.

Of course, in some physical systems closed vortices may exist. Here the flow circuits form closed loops. As long as the loops are closed no label can reach them, and they are isolated from the system. If the vortices open and close label may enter their orbit and leave. If this type of motion is rapid, the effect would average out. If it is slow, then a stable $F(t)$ will not exist and the entire method fails, since at certain times injected tracer is captured by a vortex and later released, and at other times it is completely excluded.

How, then, can label trace out a portion of the volume more than once by recirculation before its appearance at the outflow and thus give a false result? This type of error can only occur if label and fluid are separated and later recombined. In a circulation experiment, if extravascular components are included in the measured volume, then an experiment with a proper tracer will include these components since the $F(t)$ given by the tracer will yield the correct $F(t)$ for the substance being traced. Suppose, however, that radioactive Na is used to trace plasma. This is not a proper tracer experiment since a separation of tracer occurs and Na* diffuses out of the system leaving the plasma behind. On its return it is associated with a different mass of plasma which emerges later in time. Not only can label "slip" or lag behind in this way but it might in other cases appear sooner at the outflow than the substance being traced. In general any process which selectively removes the tracer from the substance being traced can produce errors of this sort.

The preceding derivation was based on the assumption that Q was constant and could be removed from under the integral sign. Actually this is unnecessary, and the preferable approach is to employ the variable $\tau = Qt/V$ which will be discussed further on pages 193–195 of this chapter. We have already noted the simplification which this variable provides in the laminar flow problem. In terms of this variable we have from eq. 19

$$\int_0^\infty \mathbf{F}(\tau)\, d\tau = \bar{\tau} = 1 \tag{22}$$

where $\mathbf{F}$ is now expressed as a function of τ. Here steady flow would not be required. Of course if flow is unsteady the $\mathbf{F}(\tau)$ which results may in some cases vary with flow rate, if the dispersion properties of the labyrinth are flow dependent.

The Shape of First-Circulation Curves: Stochastic Models

Experiment shows that, in the heart and lungs of a normal mammal, a short bolus of label introduced at the inflow generates a broad distribution in time of outflowing label. Similar effects are observed in other vascular labyrinths also. From the practical point of view, if it were possible to precisely characterize the shapes of these "first-circulation curves," the inconvenience of graphical or numerical integration would be eliminated. There would be other advantages such as the more ready detection of artifacts. From the basic point of view it would also be of value to understand the shapes and all factors governing them.

There would seem to be little hope in considering such factors as local vascular architecture in the analysis. The temporal dispersion which blood flowing in tubes may exert on a bolus of label includes both the deterministic dispersions illustrated in Fig. 55 and the possible rounding and smearing of the curve produced in practice by the tumbling of erythrocytes and other effects. In addition a profound effect will be noted in a circulatory labyrinth such as a capillary bed because of the great variation in paths which the blood takes on its passage through the maze.

A momentary consideration of the complexity of the wandering paths of the individual elements of the dye will convince us that the principal effect of the labyrinth is the production of a rather non-specific smearing of the time relations. Thus, if there is a certain amount of orderly variation of the concentration of label as a function of time in the inflowing blood, the relations will be very much less orderly in the outflow. The theoretical formulation of the problem will thus require the use of the principles of the genesis of disorder. Such an approach must lead to the basic principles of probability theory and the evolution of probability distributions.

The word "stochastic" simply means "conjectural." However, in the mathematical theory of probability the term has increasingly been applied to the analysis of random variables which change or evolve with time. This includes the theory of gambling games, of growth of bacterial colonies and other similar problems. We will thus consider the process of label dispersion as a *stochastic process.*

If we apply this concept to the progress of label through a labyrinth, provided all the label is introduced into the system at a given instant, then we may regard the number of elements emerging within a given time interval as being proportional to the probability that an element will emerge at that instant. Such an approach was taken by Sheppard and Savage (82), and was later developed by Sheppard (83, 84) and by Meier and Zierler (79). The theory of stochastic processes may be found in the literature of mathematical physics and of probability (85).

The age concept is of value in an analysis of this type. After entry into the system at zero time an element of label proceeds through an evolutionary process in which its age increases and with it the probability of its emerging at the outflow. It is inherent in the theory of stochastic processes that particles of the same age will be randomly distributed throughout the system. In the case of the bead column this distribution will tend toward the normal when represented in terms of the space coordinate x. In a certain class of stochastic processes (Markoff processes) the probability distribution representing the future history of the particle is given in terms of its stochastic state and does not depend upon how the particle reached that state. In the case of the bead column this state is characterized by the value of x. Thus, particles of zero age about to enter the system have a state $x = 0$. With increasing age, particles randomly occupy other states, but each state has its characteristic probability distribution for ultimate emergence.

Random walks versus compartments. Two types of stochastic models have been considered in an attempt to understand the shapes of

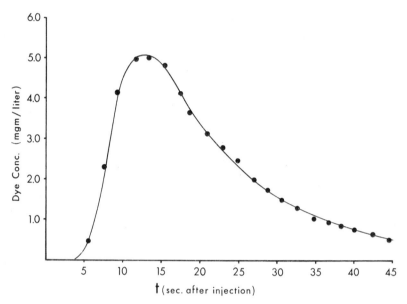

Fig. 62. Rectangular plot of experimental data obtained by Newman et al. (87) for the washing out of label from a system of serial compartments. (Data reproduced by permission of the American Heart Association.)

curves of dye concentration at the outflow of a circulatory labyrinth. In one situation, discussed by Newman and his associates (86), fluid plus label is washed through a series of stirred compartments. This "washout model" produces curves which start from zero, rise fairly rapidly to a single broad maximum, and fall less rapidly, with a quasi-exponential form, to zero (Fig. 62). By varying both the compartment sizes and the number of compartments, we can obtain considerable latitude in curve shapes. The second model is based on the use of columns of glass beads (76, 78, 84) in which the elements of injected label undergo a "random walk" among the beads on their way through the column. Neither of these models bears any valid physical resemblance to a circulatory labyrinth. However, because of certain properties of random variables, both may give a fair representation of outflow curves as a first approximation.

Random-walk models. Before considering a circulatory labyrinth, we will first discuss the basic theory of the dispersion of a bolus of label in a glass bead column. If such a bolus is introduced into a tube of glass beads, it will occupy a narrow zone in the tube. Furthermore, if the tube is very narrow, this zone may be reasonably homogeneous laterally across the tube (Fig. 63). As fluid washes through the tube, label is carried longitudinally and elements wander randomly among the beads. We may consider the distance downstream of any element of label from the point of introduction as the sum of a large number of elementary displacements occurring at random. If we plot the density of the elements of label as a function of distance downstream, the result will approach the probability that label elements experience these various displacements. We thus conceive of a probability density function of the space coordinates measured from the location where the label is initially deposited.

A   B

Fig. 63. Schematic dispersion of label in a glass-bead column showing early (A) and late (B) stages in the experiment.

In general, we may, by using dye as a label, make many of these effects directly visible experimentally in a glass-bead column. Even if there is lateral inhomogeneity initially, this condition rapidly disappears as a result of the constraining action of the tube walls and the tendency of the sidewise wandering paths to equalize the lateral concentration. At the same time the dye moves longitudinally along the column. A central maximum is retained, and moves at the same velocity as the fluid carrying the dye. The concentration also decreases everywhere; thus the dye mass broadens, and becomes increasingly dispersed. If the flow of fluid is temporarily suspended and the dye mass inspected, it appears to be symmetrical about the center of the mass. This appearance is qualitatively suggestive of the "normal curve" of mathematical statistics (Fig. 64).

That dye or other label movements of this sort should generate normal curves is a consequence of an important theorem (87). Suppose that the progress of label through the labyrinth can be expressed as a sum of a very large number of small random motions resulting from many statistically independent trials. (We avoid the concept of an infinite number of infinitesimal motions for mathematical convenience although mathematical statisticians have considered it.) In the course of the random motions, at every point along the column, an element of label may jump to other

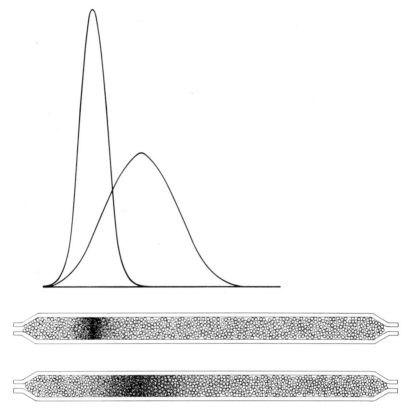

Fig. 64. Quasi-normal distribution of label at different stages during the movement of fluid through a glass-bead column.

points. We may imagine that every jump is independent of every other and also that there is a probability density function which gives the relative odds for jumps of different magnitudes, both forward and backward. Although backward jumps are permitted, most of them will be forward.

As a result of these jumps, the ultimate magnitude of the displacement of a particle of label from the origin will be expressed as the sum of n small random jumps. It is not necessary to assume that the distribution or density function is of the same form at all points along the column. Furthermore, almost any type of physically realizable function may be permitted under the theorem, provided that the individual jumps x_k are uniformly bounded, i.e., that there exists a constant A such that the absolute magnitude of x_k is less than A for all k. Suppose that the probability density function for one jump—for example, the kth—defines an expectation E_k (arithmetic mean) and a variance G_k^2 (G_k being the standard deviation). Then the "central limit theorem" asserts that

$$Pr\left(a < \frac{S_n - m_n}{s_n} < b\right) \to \Phi(b) - \Phi(a) \qquad (23)$$

In this equation, Φ is the normal distribution function; Pr is "probability that as n approaches infinity"; S_n is the sum of the individual jumps i.e., the net displacement from the origin; m_n is the sum of the expectations E_k; s_n^2 is the sum of the variances G_k^2 and n is the number of trials or jumps. As long as the processes which produce the various jumps at each trial are "physically reasonable" this theorem in effect states that as the number of jumps increases, the net displacements which they produce approach a normal probability distribution with a mean equal to the sum of the individual expected means and a standard deviation σ equal to the square root of the sum of the squares of the individual standard deviations G_k. The label should therefore be distributed along the column according to the classical bell-shaped probability curve.

In this analysis, we are assuming that there are no abrupt discontinuities in the sizes of the beads.* In particular, if there are end effects in the tube, we may expect reflection effects superimposed upon the overall label distribution. If label is introduced into one end of a finite tube, some such disturbance may occur at the entrance. However, this condition should not be readily observed at the outflow because of the smearing effect of the intervening beads. We are also assuming that the column is long in comparison with the magnitudes of the individual jumps.

Having discussed the spatial variation of label at any time in the tube, we proceed to temporal considerations. The variance measures the spread of the curve. It will increase as the number n of jumps increases, and may be expected to increase linearly with the number of trials and thus with time. Similarly, the mean of the dye distribution will increase at a uniform rate, meaning that the axis of the curve will move at a uniform speed along the tube as long as the fluid flow is uniform. The theory thus predicts a normal distribution curve moving horizontally and spreading progressively out (Fig. 64). Because of the spreading effect, label will tend to move ahead of the peak of the curve and faster than the velocity of fluid flow. This situation will cause dye to appear early at the outflow and produce a rapid upsweep in outflowing concentration. After the maximum has passed label will be increasingly delayed in its arrival at the outflow

---

* The central limit theorem requires that each jump be *independent* of previous jumps. This condition will be violated if rapid changes occur in bed properties as a function of space coordinates.

because of the further spreading effect which now acts in reverse. This prolongs the downsweep of the curve.

Of course the tube will not continue indefinitely and any label at the outflow which leaves the system cannot reenter. Our theory considers this by imposing the so-called "first traversal" condition on the particles of label. The net result of this combination of effects is that the distribution in time of concentration of label at the outflow has the form of the curve of the "one-dimensional random walk." This relation is frequently found

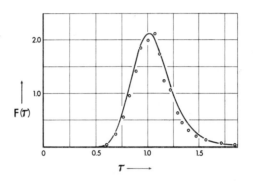

Fig. 65. Theoretical random walk curve (solid line) fitted to experimental points (circles) obtained at the outflow of a glass-bead column. (Data from Sheppard et al. (76); reproduced by permission of the American Heart Association.)

in the physical literature on kinetic theory, Brownian motion, and related subjects (85). From time to time experimental studies of dye outflow from a long thin glass bead column have been reported. Figure 65 shows one such curve (76). It is seen that agreement is fairly good between theory and experiment.

Compartmental models. The equations describing the transport of label through a catenary series of uniformly mixed compartments have been derived in eq. 15 in Chapter 5. The solution for the general case is rather cumbersome. We will therefore confine our analysis to the case of compartments of equal size, although if required, the extension to the more general case would be obvious. Here a degeneracy occurs in the differential equations necessitating a modification of the derivation. Consider the following equation which describes this system:

$$\frac{da_i}{dt} = \frac{\rho}{S}(a_{i-1} - a_i) \tag{24}$$

S is now the size of each compartment, and ρ is the transport rate through the system. If compartment 1 is initially labeled with specific activity

$a(0)$, then it can be shown by substitution that the solution for compartment i is

$$\frac{a_i(t)}{a(0)} = \frac{(\rho t/S)^{i-1} e^{-(\rho t/S)}}{(i-1)!} \tag{25}$$

This expression has the same form as the equation for Poisson statistics. Figure 66 shows a plot for a six-compartment system.

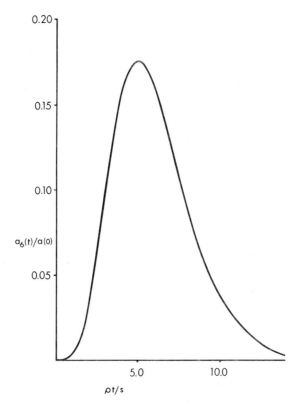

Fig. 66. Label washout from a six-compartment system. Data computed from eq. 25. Note $\rho t/S$ is the fraction of volume of a single compartment displaced, thus $\rho t/S = 5$ corresponds to $\tau = 1$.

It will be recalled that if an event has a very small probability p of occurring in a single trial the probability P_k that it will occur k times in a series of a large number n of trials is

$$P_k = \frac{(pn)^k e^{-pn}}{k!} \tag{26}$$

If label is placed in the first compartment, we can regard the washing out of label as a probabilistic process. Molecules of label make continually repeated trials to escape; the escape of label from each compartment is regarded as a success. The quantity which takes the place of the product pn is the fraction $\rho t/S$. Appearance of label in the ith compartment is regarded as $i - 1$ successes. In any one compartment, at a given instant, all particles may be regarded as being in the same stochastic state, since

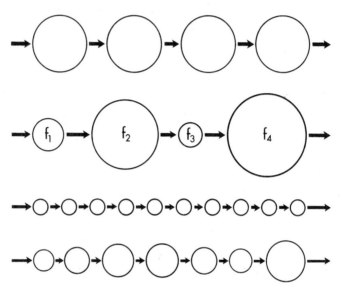

Fig. 67. Genesis of a stochastic model for label dispersion in a system of serial compartments. Top system represents a series of a few large uniform compartments. Below this is a system of variable compartments each containing a fraction f_i of the total system contents. Generalization to a stochastic system of many compartments is shown at the bottom.

they have the same probability of escape. The state is, then, a function of the compartmental index.

The recognition of the probabilistic basis of this compartmental model for the first circulation curve makes it possible to reconcile the model with that of the one-dimensional random walk. Figure 67 indicates the genesis of a compartmental system which approaches the random walk in its response as the number of compartments increases and their size decreases. Here a special case is assumed in which the probability of jumps to the left approaches zero. This situation simplifies the analysis, although a more general case can be considered where reverse transport rates are included. The analysis is also simplified by assuming uniform compartment sizes (third model from the top). Actually the central limit theorem applies equally well whether the compartments have equal sizes or whether the compartments differ provided they are small in size and large in number.*

In the same figure the situation of a few large serial compartments is also shown. When the system consisting of a large number of very small compartments is compared to the system of a small number of large compartments in their response to a bolus of label it is seen that any difference in the two situations will depend on whether or not groups of compartments, or stochastic states, can be lumped. In probability terms the difference is between a few successes for a large number of trials with small probability and very many successes for a small number of trials with moderate probability. It is well known that in the Poisson formula the probabilities approach closely those given by the Binomial law when np is finite and p is very small. At the same time, the Binomial probabilities approach closely to the normal when n is large. It is this principle which permits the outflow from a serial system of a few large compartments to conform closely to the behavior of a random-walk model. Figure 68 shows how, as the number of trials n increases, the probability of 1, 2, 3, etc., successes begins to resemble the idealized bell-shaped curve of the normal law. In other words, although the normal curve is precisely reached only in the limit $n \to \infty$ it may already be well approximated for much smaller n.

The Use of Dimensionless Variables—the τ Variable

In many physical systems there are certain "natural variables" whose use can introduce considerable simplification into theoretical expressions. In some instances variables can be combined into dimensionless quantities so that relations between them become independent of any system of units. As an illustration we may recall from Chapter 1 the expression for the specific activity in a single compartment of size S through which material is allowed to flow at a rate ρ. The differential equation is

$$S(da/dt) = -\rho a \qquad (27)$$

We introduce the variable $\tau = \rho t/S$ and substitute it into the equation, yielding

$$da/d\tau = -a \qquad (28)$$

* A word of caution is required against carrying the uniform one-way compartment system to the limit $n \to \infty$. At large n, $\kappa \to 0$ (see Appendix). The only reason for requiring n to be large is to insure that the Poisson distribution will approach the normal. This condition is sufficiently well met even for rather small values of n.

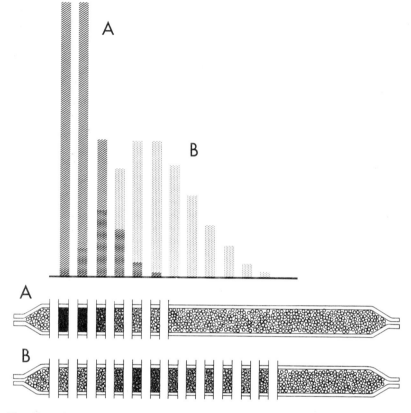

Fig. 68. Label dispersion in a lumped system showing the approach of a Poisson distribution to a normal distribution as time and the number of trials increases. The column below, in each case, corresponds to the distribution above. A: an early stage of the experiment when the expected or mean number of successes has just reached unity; B: a later stage, at which the mean number has become equal to 5, and the distribution tends toward the bell-shaped normal curve.

The equation is simplified and also its solution

$$a/a(0) = e^{-\tau} \tag{29}$$

We may also adopt the dimensionless ratio $a/a(0)$ as the independent variable. By the use of these new variables we have obtained a description of the behavior of all possible systems of this sort. This description is independent of washout rates, compartment sizes, initial labeling, and time.

The natural variable of the system is τ because it physically represents the fraction of compartmental contents which has washed through the compartment. This fraction, in turn, is a measure of the probability that label has escaped from the compartment. If no material has passed through the compartment, no label will be washed out because it is the physical movement of material through the system which causes label to be washed out. In terms of probability, τ is the fractional amount of label which has moved out of the compartment. For example, 1 per cent of the total compartmental contents washing through it will remove 1 per cent of the label (neglecting the small change of label concentration during the procedure). Similarly the ratio $a/a(0)$ is the fraction of original label which remains in the system.

There is nothing basically important in the number of milligrams of dye or microcuries of radioactive isotope in the system. On the other hand, the fraction of the initial label may be of great significance. We may recall on page 167 of this chapter the use of τ in connection with the flow of label through a cylindrical tube. We will also use τ in the next section in connection with the random walk. Other instances of the use of τ variables may also be considered.

The random walk and the τ variable. We may regard the movement of label through a labyrinth as though the label were being carried along by the medium with a progressive spreading of the label about some central abscissa. According to the central limit theorem the abscissal value of this center is the "expectation" or arithmetic mean of the horizontal components of the individual displacements of the elements. Thus the center is located at the centroid of the curve. Initially the label will cluster closely about this centroid, and the initial distribution may for mathematical convenience be represented as a delta function. This distribution then is the limit of a normal distribution having infinitesimal variance.

With progressive displacement of the arithmetic mean the axis of the distribution moves. As on page 178 of this chapter let us resolve the fluid carrying the label into a great many individual filaments of flow. In the present case we choose the filaments to have the same cross-sectional area and thus to receive the same amount of label from the initial distribution. At any point downstream after a certain interval of time t the amount of label lying in any infinitesimal element dL of longitudinal distance L will be proportional to the number of filaments having a velocity L/t. We may thus regard the arithmetic mean distance of displacement of the elements of label as t times the arithmetic mean of the velocities of the filaments and thus the mean flow rate of the fluid. Thus if label and fluid elements move together, the centroid of the label distribution will move with the mean velocity of the fluid.

It will be instructive to observe the label distribution from a moving system of coordinates which rides on the fluid and moves with the same velocity. In this case the axis of the label distribution will be stationary. As before, encounters with the beads or other dispersing elements will cause the variance about this axis to increase. For sums of random variables the variances are additive. For a glass bead column where the labyrinth has the same properties throughout, each elementary encounter will produce a displacement. Each displacement is a random variable

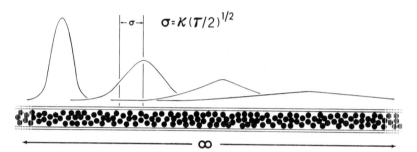

Fig. 69. Evolution of label distribution in an infinite glass-bead column.

which will then sum for the various additional encounters which occur. Encounters on the average will occur at a rate proportional to the time if flow rate is constant, and as a result the overall variance will increase linearly with time.

Of course the more basic quantity is volume displacement which can be reduced to the variable τ by appropriate choice of scale factor. The relation may in this case be specifically written as

$$2\sigma^2 = \kappa^2\tau \tag{30}$$

where σ is the standard deviation of the normal distribution of the label (see Fig. 69). The constant κ is here defined as the "randomizing constant" of the labyrinth. Where the structure of the bed is not uniform it becomes the root mean square of the κ's for the various portions of the bed weighted in proportion to the fraction of the bed with each particular value of κ. In general κ will reflect the structure of the bed in terms of its ability to produce dispersion of the label.

The overall expression for the distribution of label along the column at any given value of τ can be expressed in terms of the fraction of total label between x and $x + dx$, where x is measured from the centroid. This fraction Fdx is obtained from the well-known expression for the normal distribution

$$F = (1/\sigma\sqrt{2\pi})e^{-x^2/2\sigma^2} \tag{31}$$

If the abscissas are measured from the origin, which is the initial location of the bolus of label, the expression becomes

$$F = (1/\sigma\sqrt{2\pi})e^{-(x-\bar{x})^2/2\sigma^2} \tag{32}$$

where $\bar{x}$ is the arithmetic mean of x.

Equation 32 applies to the density of label as measured along an infinite dispersing system, but what we are interested in is an equivalent

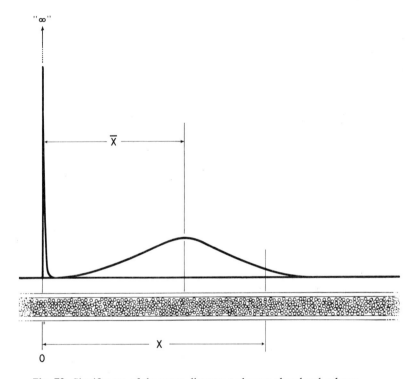

Fig. 70. Significance of the mean distance $\bar{x}$ along a glass-bead column.

expression in terms of τ variables for the case of a finite labyrinth. In obtaining this expression we may measure off from the origin a distance which includes a volume of the labyrinth equal to the ultimate system volume (see Fig. 70). If the length is taken to be x, then, as fluid is allowed to flow, $\bar{x}$ will move along the labyrinth. At any time $\bar{x}/x$ represents the ratio of the volume of fluid which has passed through to the total labyrinth volume. When the centroid of the curve has reached the point x, exactly one unit of volume has passed. Thus we may replace $\bar{x}$ by τ and x by 1 in the expression. According to eq. 30 we may insert $\kappa^2\tau$ for $2\sigma^2$. Therefore

$$F = \frac{1}{\kappa\sqrt{\pi\tau}}\, e^{-(1-\tau)^2/\kappa^2\tau} \tag{33}$$

This expression is still not final because it assumes that label at the point x represents not only material arriving at the point from the left which is about to make its first passage to points beyond, but also label which has made one passage and has just returned to x by retrograde random movement, and label making multiple passages. In an actual labyrinth it is necessary to postulate that, once the label has arrived at x, it is no longer free to wander but immediately is removed at the outflow. The situation is as though some type of sticky barrier is located at x which catches molecules of label touching it. The problem of random walks with final complete absorption involves analyzing the mathematical difference between label particles which have made one traversal of the final barrier and particles which have made multiple traversals. The analysis is considered in the Appendix. The final result is

$$F = \frac{e^{-(1-\tau)^2/\kappa^2\tau}}{\kappa\sqrt{\pi\tau^{3/2}}} \tag{34}$$

A table of this "random walk" function is given in the Appendix.

More General Random Walks

Shifted walks and double-peaked curves. Although it may be agreed that the random-walk expression gives an adequate description of the dispersion of a small bolus of label by a long thin column of glass beads, it does not apply precisely to other dispersing systems. Studies in spheroidal flasks filled with beads, for example, indicate that geometrical factors can influence the value of κ and also introduce an appreciable horizontal positive shift of the curve. The result is that an offset in the values of τ must be used to obtain an adequate experimental fit. Because of the inherent lack of resolution of systems of this sort, however, it does appear that some type of probabilistic physical model can be established which will describe the system without the requirement of precise specification of its architecture. Certainly the shifted one-dimensional random walk does seem to give a rough first approximation to the outflow curves from canine heart-lung preparations after allowance for the coronary circulation. The approximate value of $\kappa = 0.5$ is of interest (84) since the corresponding random walk curves have a fairly linear downslope in their semi-log plots (78).

The stochastic approach of course postulates that the main features which can be determined are governed by the volume of the system, flow rate, and the "randomizing constant" of the labyrinth. We should not, then, expect to learn from curve analysis a great deal more about the labyrinth than can be told from these quantities. This limitation of course applies only to experiments in which dye is injected into one end of a circulatory labyrinth and samples are obtained at the other. Certainly it may sometimes be possible to conduct indicator dilution studies in smaller portions of such a system. Injection of label into the left heart and sampling from the ventricle can of course yield more information about the heart than can be obtained from studies on the entire lesser circulation as a whole.

It is true that *gross* alterations of the labyrinth structure or of flow distribution will alter the outflow pattern. Where a large interventricular shunt occurs, a double-peaked curve is seen. In this case a rough approximation is a pair of walks in parallel. For a pair of glass bead columns in parallel (78), where each column receives its inflow in proportion to its volume, the τ values for both columns will be identical. If their κ's are the same (similar beads), the resulting curves will have the same form for both columns, differing only by a vertical scale factor. The sum will then also be identical in form to the components, and the shape of the curve for the system will be unaltered by the partition.

As the apportionment of flow between beds deviates from the ideal relation, a situation will be ultimately reached where the difference in form between the curves is great enough to yield observable departures from ideality in the summation curve. Ultimately with sufficient difference and high enough resolution in the recording system a double-peaked curve may be recognized. Whatever the partition ratio may be when dye is short-circuited, as it is in one type of cardiac septal defect, the τ relations for the by-pass and for the unby-passed circuits in the cardiovascular system should be widely different for a double peak to occur. In the detection of double peaks there may be some convenience in plotting curves on a logarithmic abscissal scale. This representation causes the curve for the lesser circulation to become quasi-normal in shape (82).

Walks in two and three dimensions. Of course we must regard the total circulation as a three-dimensional system. Situations can arise in systems of more than one dimension where the one-dimensional model still applies, which can best be seen in a simple two-dimensional problem. Consider the Brownian motion of a single particle with drift. Observing with a dark field microscope, we may start a watch at the moment the "ultra image" of the particle crosses one ocular cross-hair and time the duration of its movement until it crosses a second. The distribution of these crossing times is independent of particle movements in a direction which is parallel to the cross-hairs and perpendicular to the direction in which the projection of the particle displacement is measured. Any displacement may be described as the vector sum of mutually perpendicular displacements. The value of either of the component vectors, represented by a random variable subject to the central limit theorem, is effectively independent of the value of the other. The generalization to three or more dimensions is obvious.

In the case of the movement along a thin column of glass beads the displacement may be resolved into three mutually perpendicular displacements. One is taken along the axis of the tube, and the other two lie in the plane perpendicular to the axis. In the statistical sense random movements in this plane do not affect the movements along the axis. After a short walk in this plane the particle is soon reflected at the surface of the tube. Because multiple reflections soon create a situation in which the particle is equally likely to occupy any radial distance from the tube axis, the label concentration generally will soon tend to vary only with the distance along the axis.

Parallel systems of walks. The most attractive extension of the one-dimensional random walk in an attempt to apply it to the circulation is to consider a series of one-dimensional random walks in parallel. Each walk should have a different randomizing constant and a different scale factor relating the τ variable to the time. These walks would not necessarily represent discrete parallel flow paths but would represent different typical histories of particles wandering through the labyrinth in a more or less random fashion between the portions of the bed in which flow is rapid and the other slow-flowing portions. In their wandering paths from one point to another the elements of label would encounter different randomizing conditions. We might thus imagine the final distribution of dye concentration as a function of time to be a superposition of random-walk distributions. An analysis of the result would entail examining the resulting distribution to determine whether it could be decomposed into a series of subdistributions, each of which would correspond to a true random-walk function. Tests would then be required to determine whether the subdistributions differed significantly from one another.

It is suggested occasionally that the random walk model would apply only to the lungs because the heart chambers can be treated deterministically. Of course the analysis of the washing of label out of the heart alone, based on the assumption of complete or partially complete mixing, is a subject in itself. In studies on the entire lesser circulation as performed by classical methods any vestige of deterministic structure in the curves for the heart chambers alone are obscured by the overall smearing in the system arising from the random components.

Asymptotic expressions. It may be realized that, in terms of physics, the random-walk function approaches a limit both as $\kappa \to 0$ and $\kappa \to \infty$. The first limit corresponds to the situation where no spread occurs in the label about the mean. From the form of the function it is evident that as κ becomes large the values of τ for which $F(\tau)$ is non-zero cluster increasingly closely about $\tau = 1$. In the limit $F(\tau)$ approaches a delta function $\delta(\tau - 1)$, and this physically represents an initial spike of label which follows the fluid with no lateral dispersion and retains its initial form. In the compartmental approximation (see Appendix) the condition $n \to \infty$ is analogous to $\kappa \to \infty$. Here in the limit all traversals at the outflow are first traversals. However, this is true for all n in the one-way compartmental model. To this extent in theory at least the lumped one-way compartmental approximation is in conflict with the random-walk model. In practice for values of n which are not too small, agreement is rather good in spite of this fact.

We may also investigate the behavior of the random-walk function for large κ. From the table in the appendix it is evident that for values of $\kappa > 1$ the curve becomes increasingly skewed and the maximum occurs at values of τ considerably less than unity. Because of the factor κ^2 in the denominator of the exponent even moderately large values of κ will cause the exponent to become small. Thus

$$F(\tau) \to \frac{1}{(\sqrt{\pi}\kappa\tau^{3/2})} \tag{35}$$

except for the very smallest values of τ. From this we might predict that except for $\tau \to 0$ the function would approach zero for infinite κ, but this limit has no physical significance. Values of κ which are moderately large yield functions which are well approximated by eq. 35 except for the very smallest values of τ. For small τ we have

$$F(\tau) \to \frac{e^{-1/\kappa^2\tau}}{\sqrt{\pi}\kappa\tau^{3/2}} \tag{36}$$

As an illustration, let $\kappa = 10$. For $\tau = 0.1$, the use of eq. 35 still yields a reasonable approximation. However, with decreasing τ, as the exponent in the numerator begins to exceed unity the numerator approaches zero much more rapidly than the denominator, and in the limit $F(\tau) \to 0$. Thus $F(\tau)$ starts from the origin, rises very swiftly for small τ, and soon declines monotonically approximately as $\tau^{-3/2}$.

A semi-log plot of $\tau^{-3/2}$ may not be linear, but if data subject to biological variation are thus represented over a moderate range of τ values, an exponential representation may suggest itself. It may possibly be for this reason that some physiological data resemble the washout of label from a uniformly mixed pool (88). Of course the idea that label may wander about in a highly randomizing labyrinth so that all elements have nearly the same chance to escape in a given τ interval may be somewhat attractive as an approximate conceptual model representation of the system.

Lumping and Simulation of First-Passage Curves by Analogs

If it is granted that a few compartments of finite size provide a fair imitation of the random-walk relation, a method exists for the electronic

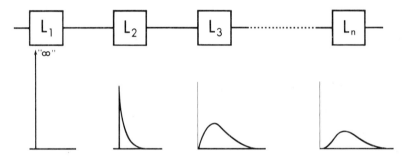

Fig. 71. Electrical analog simulation of a lumped stochastic system showing schematized curves representing voltage at the outputs of the units taken as a function of the time.

generation of curves which can be made to resemble the effects of label dispersion by the heart and lungs (89). In this respect we are again compelled by the nature of the analog method to invoke the principle of lumping, but in a serial sense. Figure 71 indicates a typical setup for an analog computer, and Figs. 72B and C shows some resulting curves obtained by pulsing the system with sharp voltage spikes. The circuit is simply a series of identical "lag units." If their time constants are sufficiently short, all voltages return to zero before the next pulse arrives. However, if this condition is not satisfied, it is necessary to provide "clampers" to discharge the units during the conditioning period prior to the arrival of the next voltage pulse.

Of course the lumping of a stochastic model into a system of a few serial compartments has no real theoretical justification when only two or three compartments are used. However, it may at times be empirically useful by the appropriate choice of compartmental parameters to approximate experimental data in this manner. Even when several compartments are employed it may be expedient to vary their parameters, particularly when integral values of n will not provide a good fit. Another device which has been employed is to set up a chain of compartments and then mix in variable amounts the outputs of the first, second, third, etc. (90).

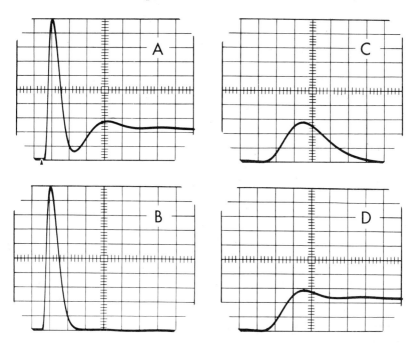

Fig. 72. Use of serially connected lag units to simulate by a lumped analogy the dispersion of label in randomizing labyrinths with and without recirculation. Ordinates are voltage (arbitrary scale), abscissas are time (5.9 msecs per div). B shows the response of five serial lag units, each having a time constant of 0.74 msec (this was to represent a single traversal of label through the lesser circulation); C shows the response after the addition of five units of time constant 2.7 msecs each, representing the systemic labyrinth. Response of a recirculating system as seen at the "aorta," indicated in A, and at the right heart in D. Photographed from cathode ray oscilloscope as in Chapter 6. Stimulus point indicated in A by the wedge in the lower left-hand corner.

In comparing the synthetic system with experimental data the moments of the curves may often be matched. We define the ith moment of the function $F(\tau)$ about the origin as

$$m_i = \int_0^\infty \tau^i F(\tau)\, d\tau \tag{37}$$

The first moment, of course, is the arithmetic mean. Table 7 yields the

Table 7

Second and third moments of $F(\tau)$. *Values are computed from expressions given in the Appendix. Moments are computed relative to* $\tau = 0$. *The first moment is unity when the* τ *variable is employed. The parameter* κ *represents the randomizing constant of a one-dimensional random walk model. For a series of identical compartments the equivalent parameter is the number* n *of compartments.*

Random-walk system				Compartmental system		
κ	Second moment	Third moment	n		Second moment	Third moment
1	1.5	3.25	2		1.5	3.0
0.707	1.25	1.94	4		1.25	1.88
0.577	1.16	1.58	6		1.16	1.54
0.5	1.12	1.42	8		1.12	1.40
0.448	1.10	1.33	10		1.10	1.32
0.366	1.07	1.22	15		1.07	1.20
0.316	1.05	1.16	20		1.05	1.16

second and third moments for the random-walk relation and for the corresponding linear "compartmental" system of n lag units. The method of derivation both for the random-walk and compartmental cases is shown in the Appendix, and expressions for m_0 through m_4 for the random-walk relation are also included. The second moments have been matched using the relation $\kappa^2 = 2/n$. The agreement between the two models is suggested in the comparison of the third moments which for complete agreement would be exactly equal. Already for $n = 4$ there is a less than 4 per cent difference.

The convolution integral.   If we are given the outflow curve from a single labyrinth which receives a small bolus of highly concentrated dye at the inflow, what will be the responses of labyrinths in various connections with one another? What will be the response of variously connected electrical analog units? The case where beds are connected in parallel is relatively clear. Here the outflow is the weighted mean of the outflows of the various tributary labyrinths. The weighting factor is the volumetric flow rate in each instance. The case of beds in series may be recognized as a more complex situation.

We may first consider a hypothetical labyrinth which receives the flow from only two parallel flow circuits with different traversal times (Fig. 73). Suppose that the inflow to these two branches is represented by a "delta function" relation. This is indicated as an alternately striped and stippled vertical bar in the figure. Then, if the two elements are chosen so small that there is no dispersion of flow within each element (indicated as two separate paths), the outflow will represent two spikes whose weight (a or b) represents the fraction of the initial spike it received, a being the striped and b the stippled component.

Let T_1 and T_2 be the transit times through the two branches while $t = 0$ is the time of introduction of the original spike. Any spike injected into

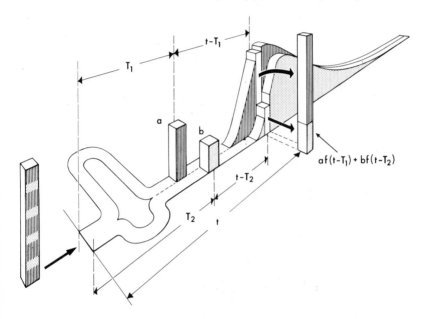

Fig. 73. The convolution principle as applied to tracer experiments (see text).

the inflow of the second bed at $t = 0$ will generate a distribution, $f(t)$, in the second bed at its outflow (shown as striped and stippled mounds in the figure*). However, the relation in the actual case for any one spike will have to be corrected for the time delay caused by the traversal time through the first portion of the system. The appropriate relations are thus $f(t - T_1)$ and $f(t - T_2)$ respectively. Therefore, the net outflow is the sum of the two distributions shifted relative to one another and with appropriate weighting factors a and b. We may conceptually remove striped and

---

* Because of practical problems in graphical representation it has not been possible to indicate from the size of the mounds that, taken together, they represent all the material in the original composite block. However, the relative differences a versus b between the two mounds have been retained.

stippled bars from the two mounds and place one above the other as shown in the figure. It is then evident that, at the outflow of the second labyrinth, the fraction of label in the small time interval at t is

$$F = af(t - T_1) + bf(t - T_2) \tag{38}$$

For a larger number n of paths in bed 1 we would have

$$F = \sum_{i=1}^{i=n} a_i f_2(t - T_i)\Delta_i \tag{39}$$

where $a_i\Delta_i$ is the fraction of label entering path i and $f_2(t)$ is the density function for traversal times through bed 2. For a very large number of divisions of label we proceed to an integral, and $a_i\Delta_i \rightarrow f_1(T)\,dT$, where $f_1(T)$ is the density function for bed 1:

$$F = \int_0^t f_1(T)f_2(t - T)\,dT \tag{40}$$

The upper limit is t instead of ∞ because the first bed cannot contribute unless label has had time to traverse it. The expression on the right-hand side of eq. 40 is known as a "convolution" and is often expressed by the symbol

$$F = f_1 * f_2 \tag{41}$$

This rather complex relation is greatly simplified by using the Laplace transform method. Let φ_1 be the transform of f_1 and φ_2 the transform of f_2, then by a well-known theorem (39) we have for the transform of F,

$$\Phi = \varphi_1\varphi_2 \tag{42}$$

The transform of the convolution is obtained by forming the product of the individual transforms of the component relations. It is this property of Laplace transforms that makes their use so advantageous in electrical circuit problems and in the study of indicator movements in the circulation.

The recirculation problem. We have already seen in Chapter 5 how a cyclically connected system can produce exponential solutions containing imaginary exponents. Although circulatory mixing may involve aperiodic as well as periodic phenomena, complex roots and oscillatory solutions are characteristic of recirculating systems. One observable situation arises when, in an intact animal or human subject, a sample of labeled erythrocytes or of plasma is injected into the right heart directly or through a close tributary. The label concentration in the arterial blood is then observed as a function of time. The theoretical analysis of this problem has been discussed previously (84, 89), but a brief résumé is included here.

We now have essentially a labyrinth in series with itself. However, we must be careful to take into account the difference between a circulation and a recirculation. If a unit spike of label is introduced into the right heart, in the ideal situation the label will be carried instantly into the circulatory labyrinth. In this idealized picture of the fate of the label the

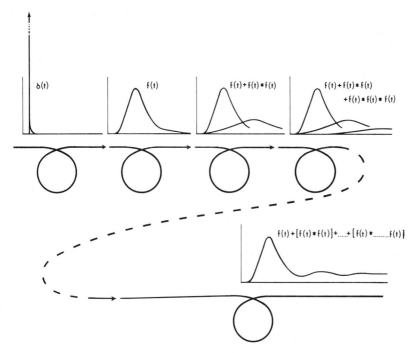

Fig. 74. Single and multiple traversals of a vascular labyrinth.

concentration in the right heart will be represented by the first circulation relation around the total circuit plus the first, second, and further re-circulations. The first circulation is taken into account by the distribution of traversal times around the total circulation. Let this distribution be described by the function $f(t)$, where t is a particular traversal time and $f(t)\,dt$ is the fraction of injected label whose traversal times lie in the interval between t and $t + dt$.

In the further analysis we must clearly distinguish between the first circulation and later recirculations. It appears at first sight that the equation governing the concentration of label $C(t)$ as a function of time will be a simple convolution of $C(t)$ on itself. Figure 74 indicates why this statement is not so. In the convolution we express the dispersing effect of a second labyrinth on the outflow from a first. If a unit spike of label is injected into the inflow of the first, the concentration at the outflow will be simply $f(t)$. This expression represents the first circulation. The second labyrinth which follows is identical to the first, but the dispersing effect is now due to recirculation and can be included as the convolution component. From this point on, any pair of beds represents a successive pair of recirculations. If we try to express the concentration of dye in the right heart as a simple convolution on itself we immediately encounter mathematical difficulties. However, it is mathematically permissible to require that the concentration on a previous circuit shall determine the effect on the concentration for the next circuit. Thus we obtain an integral equation of the form

$$C(t) = f(t) + \int_0^t C(t - T)f(T)\,dT$$

$$= f(t) + C * f$$

(43)

In addition to suggesting an equation of this type (Volterra equation) for describing the recirculation problem Stephenson (91) discussed the methods for solving it.

The meaning of eq. 43 is indicated by taking the Laplace transform of both sides. Let the transform of $C(t)$ be $\gamma(s)$ and the transform of $f(t)$ be $\Phi(s)$. Then by invoking the product law for the transforms of a convolution pair,

$$\gamma(s) = \varphi(s) + \gamma(s)\varphi(s)$$

(44)

if we solve for $\gamma(s)$, we obtain

$$\gamma(s) = \frac{\varphi(s)}{1 - \varphi(s)}$$

$$= \varphi(s) + \varphi^2(s) + \varphi^3(s) + \cdots$$

(45)

by the binomial theorem. We thus see that upon inversion

$$C(t) = f(t) + f(t) * f(t) + f(t) * f(t) * f(t) + \cdots \text{etc.}$$

(46)

The concentration in the right heart will be given by $f(t)$ plus the first, second, and further convolutions of $f(t)$ on itself. This equation is simply an expression of the fact that all dye seen in the right heart after passage of the initial bolus is that which has circulated once, plus the effect of all possible recirculations.

Use of approximate dispersion functions in recirculation problems. The serial compartment model is particularly convenient mathematically when used in a recirculating system to predict the response

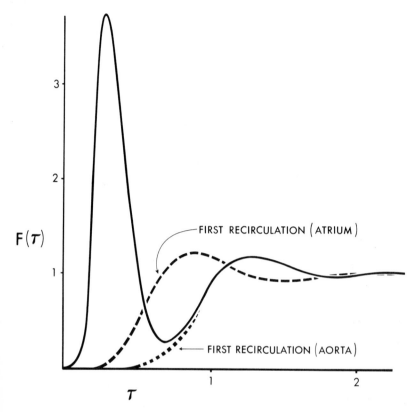

Fig. 75. Theoretical analysis of the circulatory mixing of label injected into the right heart. A serial lumped compartmental approximation was used for the dispersion of label using eight compartments for both the lesser and the systemic circulatory labyrinths. Volume of systemic circulation assumed to be three times the lesser. Solid curve represents the label concentration observed at the aorta. Dashed extrapolation represents the contribution of first recirculated (later multiply recirculated) label. The analogous curve (long dashes) is also shown for observations at the right atrium.

to a bolus of injected label. The appropriate distribution is obtained from eq. 25 by inserting $n\tau$ for $\rho t/S$ and normalizing to unit area, thus,

$$f(\tau) = \frac{n(n\tau)^{n-1}e^{-n\tau}}{n-1!} \tag{47}$$

Its Laplace transform is

$$\Phi(s) = [(s/n) + 1]^{-n}$$

Applying eq. 44, we obtain

$$\gamma(s) = \{[(s/n) + 1]^n - 1\}^{-1} \tag{48}$$

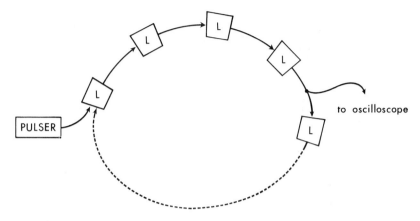

Fig. 76. Schematic setup for simulation of recirculating systems by electrical analogs.

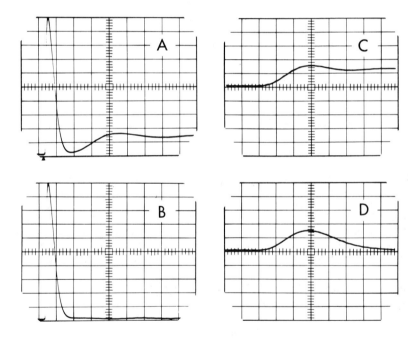

Fig. 77. Analog simulation of recirculation using the setup of Fig. 76, photographed from the cathode ray oscilloscope as in Fig. 72. See Fig. 72 for time constants of lag units. A represents observations at the aorta; C, at the right atrium less the initial "spike"; B shows the lesser circulation alone using three units ("fast" response); D shows the effect of adding seven "slow" units to the three fast ones for the systemic component of the labyrinth. Base line ($V = 0$) in C, and D is at mid-scale.

For $n = 4$,

$$\gamma(s) = (1/2)\{[(s/4) + 1]^2 - 1\}^{-1} - (1/2)\{[(s/4) + 1]^2 + 1\}^{-1}$$

The inverse, found in standard tables, is

$$C(\tau) = 2e^{-4\tau}\left[\sinh 4\tau - \sin 4\tau\right] \qquad (49)$$

The result is indicated in Fig. 75 as "first recirculation atrium." For the relations as observed in the aorta or a close tributary the expression must

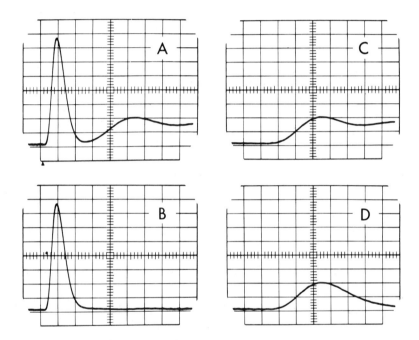

Fig. 78. Further simulation experiments. Same system as that in Fig. 77, but with four more fast units added to the "lesser" circulation. Base line is at first horizontal grid line above the base of the screen.

be subjected to a further convolution over the cardiopulmonary circuit. If the volume of this part of the circulation is taken for illustrative purposes as one third of the total volume, the results may be shown in Fig. 75 as the solid line representing first circulation and recirculation. The recirculated portion alone is indicated as a dashed line. The result is not unlike experimental curves which have frequently been obtained.

Simulation of recirculation by analogs. The approximate model used in the previous section can be readily simulated by electrical analogs (89).

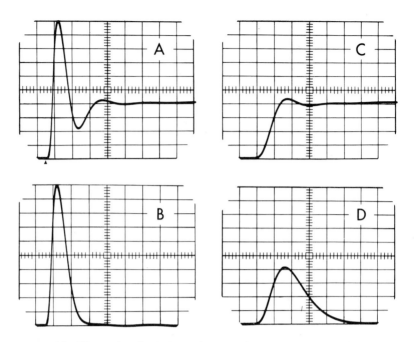

Fig. 79. Further simulations using seven fast and three slow units.

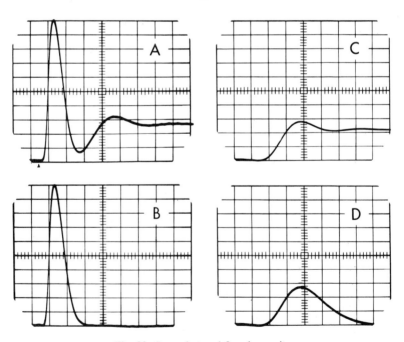

Fig. 80. Seven fast and five slow units.

The typical circuit is shown in Fig. 76, and resulting curves obtained are presented in Fig. 72 and Figs. 77 through 80. In a recirculating system, of course, label which is not lost from the circulation will ultimately be uniformly distributed with uniform concentration throughout the entire vascular pool. As seen in the figures, at the end of each sweep of the oscilloscope, the voltage does approach a constant, non-zero value. It is thus essential to equip the lag units with clampers to discharge the condensers to zero at the end of each display cycle.

Mathematical Tools for the Physiologist

Space limitations have prevented a complete coverage of the tracer method in circulatory physiology. For example, we have not discussed interesting recent theoretical research on the problem of valvular regurgitation in heart disease. An analysis based on finite-difference equations should be of considerable interest (93). We may regard this chapter as successful if it adequately demonstrates to the cardiovascular physiologist the potential value of existing tools of mathematical physics which are presently at his disposal.

10

The Movement
of Tracer Substances
from the Circulation

A tracer substance initially placed in the circulation will not remain there indefinitely. Some tracers such as injected tritium-labeled water will be lost in appreciable fraction before one circulation time is complete; others such as N^{15}-labeled hemin in newly formed erythrocytes may remain for many weeks. In each case some type of physiological information may be sought from the manner in which the concentration of label varies in the blood as a function of time. In some cases information may be obtained, and in others, contrary to initial expectation, little if any knowledge emerges from the experiments. Success or failure will depend at least in part on an understanding of the mathematical principles which affect the concentration in the circulation at various times after injection.

Disappearance of Injected Substances—General Considerations and Isolated Systems

Since the majority of experiments have been devoted to the fate of substances injected or infused into the circulation we will devote considerable space to this subject, although there is no real reason why the appearance of tracers in the blood after tissue localization should not provide a useful field of physiological research also. Many important implications may be considered in relation to the general kinetic problem of physiological translocation. We have indicated in the previous chapter how the response of a system to a "delta function" type of input and to a step input can be readily related to one another. Since in most cases, then, the response to a single injection will determine the response to other types of input we will emphasize this type of experiment.

Figure 81 shows a general type of disappearance curve for injected tracer substances. After the sudden injection of label the curve shows an initial "periodic phase" in which concentration oscillates as a function of time. This phase is rapidly damped out by dispersion of the label during the phase of rapid circulatory mixing, after which there is an "aperiodic phase" during which the label concentration declines monotonically. In

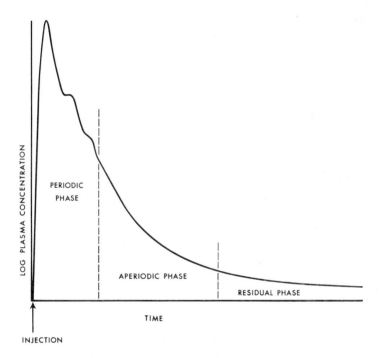

Fig. 81. General semi-log curve describing the disappearance of injected tracer from the circulation, showing periodic, aperiodic and residual phases.

many instances portions of the curve when plotted semi-logarithmically tend to be quasi-linear. Ultimately the rate of decline becomes sufficiently small that the concentration is nearly constant. This may be termed the "residual phase." Of course this curve is only schematic and in any particular situation may vary greatly depending on the substance, the system, the point and method of injection, and the point of observation. In many instances, particularly in early work, portions of such curves were not recorded because of experimental difficulties or for other reasons.

Disappearance from a uniform pool. One popular model which has been used in analyzing "transcapillary" movement of label is based on the concept that the circulation forms a uniformly mixed pool. If only a small fraction of label is lost from the blood in several circulation times it might be expected that the concentration of label would be nearly uniform everywhere in the blood. Usually experimental evidence has not been offered to establish the validity of this assumption, but because the model is a simple one it has been popular among research workers.

Irrespective of where the label is going after it leaves the blood, in this model the fraction of label lost from the circulation is obviously equal to the fraction of S which has moved out, provided that there has not been sufficient time to permit appreciable backflow of tracer. Upon following the concentration further in time, we usually note a falling off in loss rate. If the departure of the curve from linearity is not too great it is often possible to obtain a nearly linear semi-log plot of the data at least for moderate values of time. This suggests the use of a two-compartment model for analyzing the data, one a vascular compartment and one an extravascular. However, if the curve is more carefully outlined it often becomes evident that the semi-log plot is not linear and a three compartment model fits the data more precisely.

Unfortunately, because of the principle of lumping, it is possible to fit the data equally well with more than three compartments, and usually the results are not sufficiently precise to decide which multicompartment model is the best. A further objection is that non-uniform mixing of the extravascular compartments can also affect the data. At this point the investigator will usually decide that a measurement of the arithmetic mean rate of loss of the tracer from the blood early in time will suffice and that studies on the blood levels alone can provide little further information.

First-order disappearance processes. The general mathematical problem of exchange of substances between the blood and the tissues is very complex (94, 95, 96). In observing levels of a tracer substance as a function of time and of location in the circulation, it may be possible, at times, to assume for simplicity that the interaction between the circulation and its surroundings follows a first-order relation. The approximation may perhaps hold fairly well where tracer particles are removed through phagocytosis by reticuloendothelial cells. Also, events early in time during the disappearance of substances such as labeled water or K^{42} might be treated with partial success by this method since the extravascular pools are thought to be large. It is true, however, that not all physiologists welcome this assumption as holding in every case and the extent to which early backflow of tracer occurs is a matter of dispute. At the present time, at least for the circulation as a whole, the prospects for a more general mathematical treatment seem to be poor. We will therefore discuss the first-order kinetic problem even while recognizing its limitations. As an indication of the complexity of the backflow problem a discussion will be given of backflow in one highly restricted case later in this chapter (pages 219–221).

Local first-order processes; tubes of flow. It is illustrative to consider first-order processes locally in a portion of the circulation. Even though the paths of the label may be wandering, we may in some instances describe them in terms of a system of elementary "tubes of

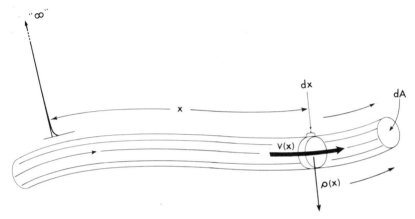

Fig. 82. Elementary tube of flow concept. An initial delta-function distribution of tracer is shown at $x = 0$. $\rho(x)$ represents the first-order rate constant for loss from the fluid element.

flow." In classical streamline flow of an incompressible Newtonian fluid the tube of flow would include a small region of space surrounded by a relatively long cylindrical surface which is formed by a sheath of adjacent streamlines (Fig. 82). In this familiar approach to fluid mechanics the cross-sectional area dA of the tube may be taken small enough that all particles within it, and situated at a given moment in a given plane perpendicular to the tube axis, have essentially the same velocity parallel to the surrounding surface. Because the surface contains a set of streamlines, no fluid will move across it and so particles of fluid which are initially situated within the tube will remain there and move parallel to the tube axis.

Let us suppose that a small volume element $dA\, dx$ is located at a distance x measured along the tube of flow from some arbitrary origin and the velocity at x is $v(x)$. A number of situations may be imagined in which the number of particles of labeled material S which disappear from $dA\, dx$ in the tube at x is proportional to the amount present. For example, material may be reacting chemically in a first-order reaction, S can exchange with non-labeled S or phagocytic processes might be removing S in particulate form. In such cases the fractional rate of loss of tracer will be $\rho(x)$. It is possible to analyze the problem even if $\rho(x)$ varies along the flow tube.* The specific activity must be considered as a function both of x and of the time t and if we are given some initial distribution along the tube $a(x, 0)$ at time zero we may desire the corresponding values $a(x, t)$ at some later time. A difference must be recognized between a considered at some fixed location at varying times and a observed as a function of x at some instant of time.

First-order processes for constant v and ρ. The most informative situation occurs when the initial distribution is a delta function or unit spike at $x = 0$ (see Fig. 82). Suppose first that the velocity is constant along the tube, as it would be if dA were uniform at all points since, with an incompressible fluid, the flow rate dQ through the tube would be constant at all points. In this case, if no tracer were lost, the initial spike, $a = \delta(t)$, would be propagated undiminished along the tube as the fluid carried it along. At any time $t = x/v$ downstream, the spike would arrive at x. The specific activity is then given by the relation $a(x, t) = \delta(t - x/v)$. However, the activity is being lost by a first-order process with a rate constant ρ which is uniform along the tube. In this case the loss of activity proceeds whether the spike moves or stands still, the relation being the usual one for first-order processes

$$a(x, t) = \delta(t - x/v)e^{-\rho t}$$
$$= \delta(t - x/v)e^{-\rho x/v} \tag{1}$$

since the delta function has no value except when $t = x/v$.

General first-order processes. As in Chapter 8 (pages 155–157), if $[c]$ is the concentration of S, we recall from vector analysis that the time rate of change of $[c]a$ is given by the negative divergence of the flux $\mathbf{J}^*$ of label. Here we are concerned only with the component of $\mathbf{J}^*$ due to bulk transport, i.e.,

$$\mathbf{J}^* = \mathbf{J}a = [c]va \tag{2}$$

taking v as the velocity. This is parallel to the flow axis.

The velocity can be a function of x but usually will not vary very much with x compared to the variation of a, since sharp changes in tube diameter

---

* If exchange at a rate ρ were occurring with very large extravascular compartments ρ would represent a true first-order rate constant. In this chapter ρ will represent a *fractional* rather than *absolute* exchange or transport rate.

are not to be expected as a general rule. Thus we may write for the divergence

$$\nabla \mathbf{J}^* = -\frac{\partial}{\partial x}\left([c]va\right) = -[c]v\frac{\partial a}{\partial x} = [c]\frac{\partial a}{\partial t} \tag{3}$$

We divide both sides by $[c]$ and subtract the first-order fractional disappearance rate term $\rho(x)a$, yielding

$$\frac{\partial a}{\partial t} = -\rho(x)a - v(x)\frac{\partial a}{\partial x} \tag{4}$$

It can be verified by differentiation that

$$a(x, t) = a_0(t - T)e^{-\rho_m T} \tag{5}$$

is a solution of eq. 4. We define $T = \displaystyle\int_0^x \frac{dx}{v(x)}$ as the time required for fluid originally at the origin to arrive downstream at x. Also,

$$\rho_m = \frac{1}{T}\int_0^x \frac{\rho(x)\,dx}{v(x)}$$

is the mean value of ρ taken over the path from the origin to x. The initial value $a_0(t)$ is the specific activity observed at the origin as a function of time. We have not proved that eq. 5 yields the most general solution. However, from the physical point of view it contains all the requirements of such a solution. The time delay T between 0 and x is correctly given as the sum of all $dt = dx/v$ taken over the path. The attenuation factor $\rho_m T$ is correctly given by the sum of all attenuations $\rho\, dT$ taken over the same set of variables.

Exchange between a vessel and its surroundings. As soon as backflow of tracer from extravascular pools is considered the problem becomes extremely complex. A first point of departure involves a highly artificial model showing exchange between an elementary tube of flow and its surroundings (97, 98). Tracer is considered to move only radially and to return to the tube of flow from an extravascular pool which is uniformly mixed in a direction perpendicular to the flow axis. Of course this model is only distantly related to physiological reality. The mathematical treatment takes its origin from the theory of heat exchangers. Consider a system similar to Fig. 83. In the present case ρ represents the rate of *exchange* of tracer with the external pool as a fraction of the *internal* pool per unit time. Similarly we define ρ' as the equivalent fraction of the *external* pool exchanged per unit time. We assume that ρ, ρ', and v are constant. If the internal specific activity is $a(x, t)$ and the external $a'(x, t)$

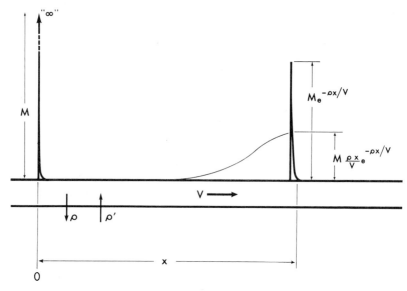

Fig. 83. Hypothetical system used for approximate theoretical study of tracer exchange allowing for back flow of tracer. Initial delta function M is attenuated exponentially. Back flow is represented by a tail following the attenuated "spike."

we may write the differential equations as

$$\frac{\partial a'}{\partial t} = \rho'(a - a')$$

$$\frac{\partial a}{\partial t} = \rho(a' - a) - \frac{v}{dx}\frac{\partial a}{} \tag{6}$$

Solution has been obtained by Sangren using Laplace transforms (97). We define $\alpha(s)$ as the transform of $a(x, t)$ taken with respect to t and $\alpha'(s)$ as the transform of $a'(x, t)$. Let $a(x, 0)$ be the initial distribution of specific activity along the system. Thus

$$s\alpha' = \rho'(\alpha - \alpha')$$

$$s\alpha - a(x, 0) = \rho(\alpha' - \alpha) - \frac{v}{\partial x}\frac{\partial \alpha}{} \tag{7}$$

Eliminating α', we obtain

$$\frac{\partial \alpha}{\partial x} + c\alpha = \frac{a(x, 0)}{v}$$

where

$$c = \left(s + \rho - \frac{\rho\rho'}{s + \rho'}\right)\frac{1}{v} \tag{8}$$

A fairly general treatment of the problem has been outlined for arbitrary $a(x, 0)$, but a sufficiently illustrative case is provided for an initial delta function distribution at $t = 0$, taken as a function of time rather than of x.

The solution of the equation in α is

$$\alpha(x, s) = e^{-cx}\left[\int_0^x \frac{a(z, 0)}{v} e^{cz} \, dz + A\right] \tag{9}$$

Because α must be bounded for all values of x, because the real part of c must be positive, and because $a(x, 0) = 0$ for $x < 0$, A must be zero. Thus

$$\alpha(x, s) = e^{-cx}\int_0^x \frac{a(z, 0)e^{cz}}{v} \, dz \tag{10}$$

When $a(z, 0)$ is a delta function with area M under the initial spike (where concentration is a function of time),

$$\alpha(x, s) = Me^{-cx} \tag{11}$$

By a series of inversion relations (97) we obtain

$$a(x, t) = Me^{-\rho x/v}\Big(\delta(t - x/v) + e^{-\rho'(t - x/v)}$$

$$\times \left\{I_1\left[2\sqrt{\rho\frac{\rho'x}{v}\left(t - \frac{x}{v}\right)}\right]\sqrt{\rho\frac{\rho'x}{v(t - x/v)}}\right\}\Big) \tag{12}$$

In performing this inversion we note that the inverse of $e^{b/s}$ is given by $\delta(t) + I_1(2\sqrt{bt})\sqrt{b/t}$ where $\delta(t)$ is a unit delta function at $t = 0$. The function I_1 is a standard Bessel function of imaginary argument, and b is any quantity which is not a function of t. Since time enters $a(x, t)$ in terms of $t - x/v$ the distribution moves along the vessel with uniform velocity v. The first term on the right-hand of eq. 12 represents the original spike which is attenuated in the same manner as in eq. 1. The second term represents a tail which follows the spike (Fig. 83) and is the result of label moving back from the external compartment, exchanging perhaps more than once across the boundary of the tube. A graph of the tail function is shown in Fig. 84. In drawing the graph we adopt the "natural" coordinates $\rho x/v = p$, $\rho'(t - x/v) = q$. In these units we have

$$a(p, q) = Me^{-p}\{\delta(q) + e^{-q}[I_1 2\sqrt{pq}]\sqrt{p/q}\} \tag{13}$$

M is the area under the initial spike, now taken as a function of q instead of t. (Note that eq. 13 differs from the corresponding equation given by Sangren and Sheppard (97), which contains an error.)

Arterio-venous differences and non-uniform mixing. When a tracer substance is lost from the circulation in appreciable amount during one mean circulation time the uniformly mixed pool assumption is no longer valid. Whether investigation now is directed to studies of isolated organs or whether we observe all the features of disappearance curves

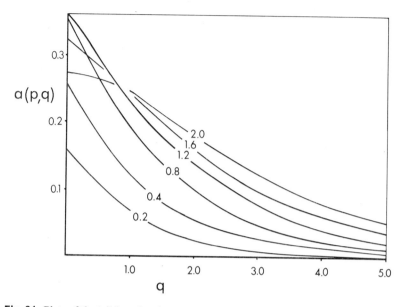

Fig. 84. Plots of the tail function (see eq. 13 in this Chapter). Numerical values of p are given on each curve. Ordinates are values of $a(p, q)$, abscissas are values of q.

recorded in intact animals, we encounter characteristic difficulties in each case in interpreting the results. The unphysiological isolated organ system is only simpler in a relative sense. Under experimental conditions the much more complex intact animal yields only a somewhat better approximation to normal physiological behavior. Improved physiological knowledge may be expected to come from a coordination of the results from both approaches.

Where an isolated organ system is perfused from a large stirred reservoir with recirculation in a closed system the concentration of label may sometimes be recorded both at the inflow and the outflow as a function of time. The rate of decline of concentration in the reservoir together with a knowledge of the system volume can be combined to obtain the rate of loss of tracer from the system provided that the reservoir is large enough. If this rate is $-dS/dt$, and if "arterial" and "venous" concentrations are C_a and C_v respectively, the flow rate Q may be related by the equation

$$\frac{dS}{dt} = Q(C_a - C_v) \tag{14}$$

This tracer relation is employed by physiologists in determining Q by the Fick principle. Of course the method is not confined solely to isolated organs since cardiac output is traditionally determined using oxygen, for example, as a "tracer" and a spirometer as the "reservoir."

Experiments involving perfusion from a reservoir are usually intentionally designed so that the loss of tracer from the system occurs entirely in the biological portion of the circuit. Then $-dS/dt$ represents the accumulation of tracer into the organ or organs under study. The equation is completely general and represents a simple balance between input into and outgo from the system. By integration of the relation it is possible to establish the total amount of tracer in this portion of the system, and if initially C_a and C_v are zero we have for the organ content of label

$$S_o(t) = \int_0^t Q(C_a - C_v)\, dt \qquad (15)$$

In this integration it is not required that Q, C_a, or C_v be constant. In particular C_a and C_v may vary where a bolus is injected into the system and allowed to equilibrate. If the experiment is continued until equilibrium is established with a tracer which remains within the vascular system of the organ, both C_a and C_v will approach a common constant value C_{eq} and S_o will approach S_{oeq}. If the equilibrium concentration C_{eq} exists, it represents the concentration of label within the organ being perfused. At equilibrium $S_{oeq} = C_{eq}V$, where V is the system volume and

$$V = (1/C_{eq}) \int_0^\infty Q(C_a - C_v)\, dt \qquad (16)$$

This method succeeds ideally where the arterio-venous difference of label may be directly determined. In other instances, it may be necessary to subtract independent determinations, one from the other. This may, through systematic error, introduce considerable uncertainty into the value of the integral, the error becoming greater as the upper limit of the integral increases.

Of course the Fick principle is based on the existence of a relationship between dS/dt and flow rate. Usually a tracer is selected which is lost out of the blood, producing a resultant large arterio-venous concentration difference. Ideally this difference should be relatively independent of system parameters. In the case of organ perfusion with such tracers, on the other hand, we may be interested in the relation between dS/dt and properties of the vascular labyrinth. Here a relation between dS/dt and Q may be fortuitous and undesirable though it may be found in many reservoir perfusion experiments since tracer cannot disappear from the system at a rate greater than it is being delivered to the organ being perfused. Perfusion at varying flow rates and extrapolation to infinite rate may be required to obtain the desired information.

"Diffusible" versus "indiffusible" tracers. In some instances a system may be perfused with an "indiffusible" tracer which is known to remain within the vascular tree. If a bolus is introduced at the inflow, the outflow curve will yield information concerning the distribution of transit times through the circulatory labyrinth. These considerations were discussed in Chapter 9. When the tracer is "diffusible," loss of material out of the circulation will tend to prolong the outflow curve. In a double-tracer experiment if both tracers are introduced simultaneously, the outflow concentration of diffusible tracer will tend to remain below that of the indiffusible initially, since loss is occurring from the blood into the extravascular spaces. By integration of arterio-venous differences it may then be verified that a greater amount of diffusible tracer is contained in the organ. If the experiment is continued for a sufficiently long period, back-flow into the circulation may eventually be observed. In some instances most if not all of the lost material may then be recovered.

It may be useful in that case to consider the significance of the system volume which may be estimated from the mean transit time of the *diffusible* tracer. Since the diffusible tracer appears initially in less amount than the indiffusible, and later in greater amount, the centroid will be displaced toward larger values of time yielding a larger volume. If a stable pattern of extravascular flow paths existed, the volume derived from these data might represent an anatomically characteristic extravascular volume. Usually, however, the tail of the curve is greatly prolonged and merges into background "noise." Difficulty will then occur in deciding when to conclude the experiment. Centroids determined by integrating over larger and larger values of time become progressively greater. This extravascular space may gradually approach the entire anatomical space but probably rarely reaches it. Thus volumes determined by this method should usually remain less than volumes determined by perfusing the organ to equilibrium in a recirculating system.

Initially the difference between the area under the curve for the diffusible and indiffusible tracers represents label predominantly going out of the blood. If Fig. 83 has any validity in the practical sense this phase represents primarily the progressive attenuation and smearing of the initial spike. Late in time the backflow represents a fusion of all the tails of the individual tracer elements. Because of the smearing effect, however, it is not possible in general to determine how soon backflow from the more rapidly equilibrating elements occurs. Late in time there may be portions of the labyrinth where tracer is predominantly moving out and other portions where it is predominantly returning. For this reason it is not possible to consider that the point where the curves for diffusible and indiffusible tracers intersect represents the point of equilibration of diffusible tracer with the extravascular pool.

The system in Fig. 83 assumes a concentration of tracer which is externally constant, irrespective of the radial distance from the axis of the vessel. Actually for very small vessels the volume of the region with which rapid equilibration occurs may be quite small, but this minute cuff represents only a small portion of the cylindrical region surrounding the vessel. Because the volume of the larger sleeves surrounding the vessel increases rapidly with radial distance these more distant regions may equilibrate slowly. The situation becomes even more complex in actual tissues where cell boundaries soon intervene.

Outflow versus inflow in a perfused organ.   We will now consider the relation between the concentration or specific activity of a tracer at the outflow and at the inflow of a vascular labyrinth when exchange can occur with large extravascular pools (99, 100). In Chapter 9 we proceeded to generalize the histories of label transit through individual flow paths in the system. A probability density function was invoked by which the individual elements contributed their fraction of label at the outflow in proportion to the fraction $F(t)\,dt$ having transit times between t and $t + dt$. We can follow a similar course once more, but now the individual flow circuits differ not only in transit time but in mean exchange rate. Furthermore if backflow is postulated, the ratio of external to internal pool size must be included even for the approximate model of Fig. 83. Because such detailed models are mathematically difficult and because simpler models already yield interesting information we will consider only the problem of first-order kinetics without appreciable backflow.

Let the inflowing specific activity relation be $a_0(t)$. For a single fluid element eq. 5 gives, for the outflow at a point x downstream,

$$a(x, t) = a_0(t - T)e^{-\rho_m T}$$

where T and ρ_m are the transit time and the mean exchange rate respectively. We thus obtain the specific activity for a collection of elements as the weighted mean for the individuals, using as weighting factors the functions $F(T)$ for transit times and $\Phi(\rho_m)$ for exchange rates. As before, these functions are normalized to unit area. Finally, then, if the time relation at the inflow is $a_1(t)$ we obtain at the outflow

$$a_2(t) = \int_0^\infty \left\{ \int_0^t F(T)\Phi(\rho_m)a_1(t - T)e^{-\rho_m T}dT \right\} d\rho_m \qquad (17)$$

As in the non-diffusible tracer the upper limit on T is set by the elapsed time during which label may arrive at the outflow. In this analysis it is assumed for simplicity that $F(T)$ and $\Phi(\rho_m)$ are independent. In this case a_2 is not uniquely related to a_1 until sufficient time has elapsed for the downstream events to be influenced by those upstream. If, meanwhile, there is no tracer in the system eq. 17 will apply, but if intervening tracer is present an additional term must be included to take it into account. Usually the analysis can be arranged so that the initial level of tracer is negligible. Further considerations in the derivation of this and associated relations are found in the original paper (99).

The Circulation as a Whole

Integral equation formulation. Subject to the initial condition already mentioned, eq. 17 should hold for any "well-behaved" input function $a_1(t)$ and physiologically reasonable functions F and Φ, where Φ is the distribution of ρ's. As in Chapter 9, we may proceed to apply it to a recirculating system initially devoid of tracer. We will consider the response $a(t)$ of the system to an injected bolus of label represented by a "delta function" of initial area $a(0)\,dt$. For simplification we will consider the response which would be observed at the site of injection. By further application of eq. 17, we could obtain the response at some point further downstream. Thus we have

$$a(t) = a(0)F(t)\,dt\int_0^\infty \Phi(\rho_m)e^{-\rho_m t}\,d\rho_m$$
$$+ \int_0^\infty \int_0^t F(T)\Phi(\rho_m)a(t-T)e^{-\rho_m T}\,dT\,d\rho_m \quad (18)$$

Here, of course, the ρ_m are mean rates for the flow circuits of the entire circulation. For the non-diffusible case all values of ρ_m are zero. Thus, as in Chapter 9 we may write

$$A(t) = a(0)F(t)\,dt + \int_0^t F(T)A(t-T)\,dT \quad (19)$$

reserving in this case $A(t)$ for the non-diffusible tracer.

Equation 18 may be thrown into the form of eq. 19 by the substitution

$$a(t) = \psi(t)e^{-(\hat\rho+\alpha/\bar T)t} \quad (20)$$

yielding

$$\psi(t) = a(0)G(t)e^{\alpha t/\bar T}\,dt + \int_0^t G(T)\psi(t-T)e^{\alpha T/\bar T}\,dT \quad (21)$$

Here $\hat\rho$ and $\bar T$ are grand arithmetic means of the exchange rates and transit times for the entire circulation. $G(t)$ satisfies the relations

$$G(t) = F(t)\int_{-\hat\rho}^\infty \Phi(k)e^{-kt}\,dk,$$

and
$$\int_0^\infty G(T)e^{\alpha T/\bar T} = 1 \quad (22)$$

where $k = \rho_m - \hat\rho$, and the second relation defines α.

It is instructive to consider the significance of these expressions. The parameter α represents the effect of circulation rate on the kinetics of transcapillary disappearance of tracer. Since the perfusion rate affects the kinetics in the case of organ perfusion from a reservoir it may be expected that such an effect would also exist in a recirculating system. However, it may be shown that the effect will be small unless there is a rather broad spread in the distribution of ρ_m. With varying circulation rate the "arterio-venous" differences tend to adjust themselves to compensate for the rate changes (6). An effect of the exchange processes on the mixing kinetics may also be predicted since in general the "kernel" $G(t)e^{\alpha t/\overline{T}}$ will differ from $F(t)$.

When the ρ values tend to be homogeneous and the rates of transcapillary movement of tracer are small compared to the circulation rate it is evident that $\alpha \to 0$, $G(t) \to F(t)$, and $\psi(t) \to A(t)$. In this case we may represent the specific activity relation for the "diffusible" tracer as the product of the "indiffusible" specific activity and a single exponential factor whose rate constant is the overall grand arithmetic mean $\hat{\rho}$ for the entire circulation. This is in accord with our intuitive picture of the uniform pool approximation discussed on pages 215–216 of this chapter. With moderate increase in the interaction between exchange and mixing kinetics, the theory predicts that mixing oscillations will still be observed, with a simultaneous quasi-exponential attenuation during the initial phases of mixing. Finally as the ρ values become larger considerable distortion of the oscillations is to be expected until ultimately recirculation becomes so slight that little more than one single wave should appear to the gross observer. In this limiting case scarcely any vestige of a single exponential component may be predicted.

Partition of cardiac output. Although the influence of flow rate on *circulatory* disappearance curves may represent a second-order effect, in many instances the localization of tracer in *external* pools is strongly affected by circulation rate. One particular aspect of this problem is indicated in the tendency for initial localization in different branches of the arterial circulation to be determined by the fraction of the cardiac output perfusing them (101, 102). Of course if all of the tracer is removed from the blood on one pass through an organ, the rate of accumulation will depend directly on the rate of blood flow into the organ. This is the basis, for example, of the use of sulfobromphthalein or of radiocolloids for estimation of splanchnic blood flow. The rate of appearance in the organ may be determined from the rate of loss from the overall circulation since little tracer is lost into other depots except the liver (and spleen in the case of colloids). In this instance, of course, the circulation rate effect predominates because there is a large spread in the ρ values. Those for the liver approach infinity, and those for the remainder of the circulation approach zero. Although ρ does not represent an exchange rate it may be taken as a type of first-order rate constant for phagocytosis or for elimination of dye.

Recently considerable interest has been noted in the use of K^{42} and other tracers for estimating the partition of cardiac output among the different branches of the aortic outflow (101, 102). Here, again, the ideal tracer for such a purpose would be one which is lost completely into extravascular spaces on one pass through all organs receiving blood from a particular aortic tributary. Upon injection of the label into the heart the partition of label among the tissues would be in proportion to the fraction of the cardiac output which they receive. Actually in the rat (102) the extraction of K^{42} is not 100 per cent, but a lesser criterion will suffice.

Assume that the tissues of the ith branch of the arterial tree may be adequately sampled (a difficult assumption perhaps) and that at time t they contain tracer in amount $S_i(t)$. If Q_i is the rate of blood flow through the branch then, given $C_{ai}(t)$ and $C_{vi}(t)$ for the inflowing and outflowing concentration of tracer, we obtain

$$S_i(t) = Q_i \int_0^t (C_{ai} - C_{vi})\, dt = Q_i \Lambda_i(t) \qquad (23)$$

The ith function $\Lambda_i(t)$ may be considered as a localization function for the ith branch. If we can assume adequate mixing of label in the aortic outflow, $C_{ai}(t)$ will be identical for all branches. A similar assumption is questionable for $C_{vi}(t)$, and, in fact, in the rat there is evidence that it does not hold for certain tissues. If the differences are not great and if the $C_{vi}(t)$ are relatively small compared to $C_{ai}(t)$, there will be a tendency for the $\Lambda_i(t)$ to approach a common value $\Lambda(t)$. If this is the case, then, approximately

$$\frac{Q_1}{S_1} = \frac{Q_2}{S_2} = \frac{Q_i}{S_i} = \frac{1}{\Lambda} \qquad (24)$$

Under such idealized circumstances each "organ" will receive its particular share of label in proportion to its fraction of the cardiac output, and because of the similarity to the other organs in its outflow pattern it will lose a similar fractional share. Under these conditions tissues will not tend to gain label from others or to lose label, and a stationary tissue level of tracer may be expected under the idealized assumptions of this theory. To the extent that the idealization breaks down, a gradual redistribution of label is predicted.

Analog simulation of tracer disappearance. The movement of tracer into extravascular pools whether it is by phagocytosis, by transcapillary exchange, by metabolism, or by other mechanisms will not, in general, be a localized process, but must occur diffusely at various points in the circulation. For this reason the use of lumped models may be somewhat hazardous. Since the conventional analog computer is essentially a lumped type of machine, its possible use in simulating tracer disappearance will depend upon the best utilization of the art of lumping of distributed systems. We may be guided to some extent by the techniques used by engineers in the lumping of such systems as heat exchangers (62). The lumped system will give a better approximation as the lumping is more widely distributed over its elements. Thus a computing system having a large number of operational amplifiers will often be required.

To illustrate the semi-empirical technique, we return to eq. 17 relating the outflow and inflow characteristics of a vascular labyrinth where tracer is lost by first-order processes. For simplicity we will consider the response at the outflow to a delta function input at zero initial time. Here, since $a_1(t)$ has no values except for zero argument, $T = t$. Replacing T by t and removing the integral sign, we have

$$a_2(t) = F(t)\int_0^\infty \Phi(\rho_m)e^{-\rho_m t}\,d\rho_m$$

$$= F(t)e^{-\bar\rho t}\int_{-\bar\rho}^\infty \Phi(\rho_m)e^{-\Delta\rho t}\,d\rho_m \qquad (25)$$

where $\bar\rho$ is the arithmetic mean of ρ_m and $\Delta\rho = \rho_m - \bar\rho$ is the deviation of ρ_m from the mean. We may now define the ith moment of $\Phi(\rho_m)$ about the mean as

$$M_i = \int_{-\bar\rho}^\infty (\Delta\rho)^i\Phi(\rho_m)\,d\rho_m \qquad (26)$$

Thus

$$a_2(t) = F(t)e^{-\bar\rho t}\int_{-\bar\rho}^\infty \Phi(\rho_m)\left[1 - \Delta\rho t + (\Delta\rho)^2\frac{t^2}{2!}\cdots\right]d\rho_m$$

$$= F(t)e^{-\bar\rho t}\left[1 + \sum_{i=2}^\infty (-t)^i\frac{M_i}{i!}\right] \qquad (27)$$

The first term in the sum is unity since Φ is normalized, the second is zero since the moments are taken about the arithmetic mean. If in their distribution, the values of ρ_m cluster about the mean, the higher moments of most physically "reasonable" functions will tend to be small and the factorial in each denominator will be large. For this reason, particularly if $F(t)$ does not include large t values, if the rates ρ_m are not too great, the attenuation of $F(t)$ will be adequately described by a single exponential factor $e^{-\bar\rho t}$. If $F(t)$ can be simulated by a series of lumped RC units as we have described in Chapter 9 it may then be permissible to connect resistances from the condensers of the lag units to ground to simulate loss of tracer to infinite extravascular pools by first-order processes. As time increases, the deviation of the attenuation factor from a single exponential begins to become more evident. From eq. 27 the deviation first manifests itself in the t^2 term through the second moment M_2. By combining two parallel

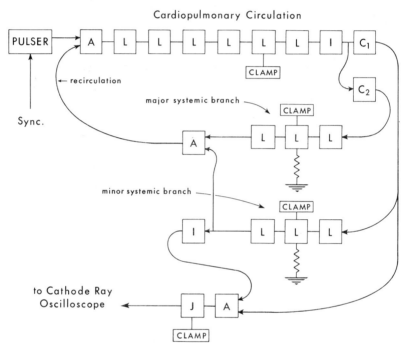

Fig. 85. Lumped electrical analog model for simulating "cardiac output partition" experiments. See text for description. Designation of the operational units is the same as in Chapter 6.

elements with two exponential constants chosen to match the value of M_2 this effect may be simulated to a fair approximation. For greater deviations from exponential behavior further branching may be required, but the complexity of the electrical network increases rapidly as further precision is required.

Another type of deviation from exponential behavior will occur if back-flow of tracer from the peripheral pools is to be simulated. Here again, if the pools are large and fairly homogeneously distributed, it may be possible to introduce capacitance elements in series with the resistive bypass connections to ground. If after varying the arrangement of connections the result does not appear to be sensitive to the choice of elements and their connections it may be possible to obtain a satisfactory approximate simulation. The hazards of such experiments are obvious, and they should not be undertaken without an adequate understanding of their mathematical and physiological basis.

Figure 85 shows a lumped model which has been employed in our laboratory to study the effect of varying system parameters on the cardiac output partition. The dispersing effect of the lesser circulation was simulated by six small "lag" units L, and the systemic circulation was separated into a major and a minor branch, each simulated by three large L units. As indicated in the figure, loss of tracer was simulated by a resistor draining the capacitance element of the central unit of each branch. To reduce all voltages to zero at the end of each display cycle, clamping units were connected to some of the L units. Partition between the major and minor branches was simulated by the coefficient units C_1 and C_2 preceded by an inverter I to correct for the inversion produced by the C units. The output of the major and minor branches was recombined by an adder A. Its inverted signal becomes erect once more in the second adder which combines the recirculated signal with the initial (inverted) pulse. At the lower portion of the diagram the difference between input and output for the minor branch is integrated by element J to obtain a representation of the label content of this simulated "organ," as a function of time. A display of the output of the J unit provides some estimate of the time variation of the organ content when the input "concentration" $C_a(t)$ and the outputs $C_{v1}(t)$ and $C_{v2}(t)$ are altered.

Disappearance of Labeled Formed Elements from the Blood

The kinetics of the appearance and disappearance of formed elements from the blood has been investigated by tracer methods by physiologists for some time. The earlier antigenic technique of Ashby can be regarded as a type of tracer method for investigating erythrocyte kinetics, but the development of isotopic labeling gave considerable impetus to research in this field. The need for an adequate mathematical analysis was soon evident, and the reader will find more comprehensive details in existing literature reviews (103). Using erythrocytes labeled with a tracer which is tightly adherent, we can by following the concentration of the label in the blood determine continuously their circulating level. Upon cell death (provided there is no reutilization of the label) both cells and label are lost from the circulation. The death and replacement of erythrocytes may thus be investigated.

From the kinetical point of view we are again concerned with a stochastic process characterized by a probability density function which gives the chance that cell death will occur at any given instant of its life span. The level of tracer in the circulation following the sudden labeling of new cells at $t = 0$ may be regarded in terms of a probability distribution representing

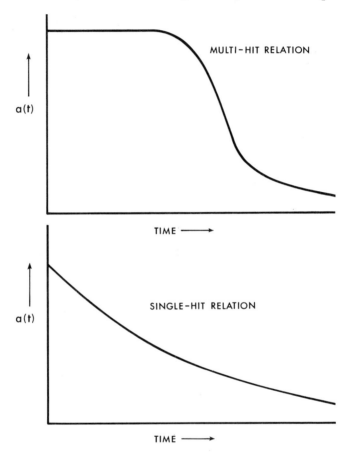

Fig. 86. Multi-hit (top) versus single-hit (bottom) distributions of blood cell lifetimes. Ordinates represent percentage remaining after time t (abscissas).

the fraction of cells whose lifetimes are equal to or greater than the time between birth and observation. If the duration of the labeling process is finite it may be necessary to consider a second probability function representing the fraction of cells produced within a given time interval. The overall level, then, will be expressible as a convolution of one function on the other. The nature of the "birth" function will depend upon the manner in which the label is handled by the organism on its way into the cells, but usually it is possible to minimize the duration of labeling to compensate for these uncertainties.

The kinetics of senescence and death of cells will depend to a considerable extent upon the factors involved in their mortality. Normal erythrocytes in several mammalian species have lifetimes of the order of several months, and under these conditions the distribution function tends toward the normal or Gaussian distribution which may be expected from the central-limit theorem. Cell lifetime is thus a random variable, and the death of a cell is regarded as a cumulation of a large number of small insults. In abnormal instances pathological factors may intervene so that cells can die as a result of sudden catastrophe. The probability distribution may then at times approach the exponential function. As in radiobiological death we may then recognize "single-hit" deaths and deaths of higher multiplicity which, with decreasing pathology, merge into the normal aging process (Fig. 86). Although the Gaussian distribution model seems to suffice, for erythrocytes other stochastic models for general senescence may be found in the literature (104). Theoretical analyses have also been applied to the study of white-cell genesis and mortality in health and disease (105, 106).

11

Matrix Methods and Fitting Models to Data

It will have become abundantly clear after reading the preceding chapters that the success or failure of the tracer method in the more complex systems depends critically upon our ability to get maximum information out of laboratory data. Given a set of perfect data with zero fluctuation observed over a sufficiently wide range of time in all compartments of a steady-state multi-compartment system, we have shown in Chapter 3 how the transport rates may be obtained from these measurements. Actually, of course, there will always be experimental error, and practical considerations will often exclude certain portions of the system from inspection or limit the number of repetitions of the experiments. If we cannot fit the data to a single unique model, is it still possible to set limits among the possible compatible models? Given data subject to experimental fluctuation, how can we obtain the maximum amount of information from them? In obtaining partial answers to these questions we must resort to more sophisticated mathematics than we have used up to this point. In so doing we will be bound by neither elegance nor rigor. Some of the missing elements of a proper treatment may be found in more advanced mathematical texts (107).*

Matrix Formulation of Tracer Kinetics

Numbers, vectors, and matrices. In previous chapters we have considered quantities such as specific activities and time as individual numbers (scalar quantities) related to the system compartments by a series

---

* A recent paper by Bergner (108) formulates the theory of tracer kinetics using the terminology and notation of abstract set theory. This approach would seem to represent the maximum sophistication which has been reached in the literature at this time.

of indices. The burden of notation thus required is usually not severe since many tracer kinetic experiments are conducted in systems of two or three compartments only. For a two-compartment system specific activities are given in the two compartments as functions of time t. They may be separately represented as $a_1(t)$ and $a_2(t)$, but this representation can be facilitated if we adopt the convention that the a's are to be represented as horizontal and vertical components of a vector $\mathbf{a}$ in two-dimensional space.

If, for convenience, the tail of the vector is placed at the origin of coordinates, the movement of the tip of the vector will trace a path in space which represents the history $\mathbf{a}(t)$ of the system as time progresses. The vector $\mathbf{a}$ is a function of the scalar quantity t, and $\mathbf{a}(0)$ will be the initial value of $\mathbf{a}$. For an n-compartment system we will consider that $\mathbf{a}$ is a vector in n-dimensional hyperspace. It is a simple matter to express the derivative $d\mathbf{a}/dt$ as the quotient of the change $\Delta\mathbf{a}$ in the vector divided by the small associated increment Δt in time taken as a limit as $\Delta t \rightarrow 0$. In this manner the derivatives of the n components may be represented vectorially, and thus the rates of change da_i/dt of specific activities in the various compartments. Of course $\Delta\mathbf{a}$ can be a change in magnitude of $\mathbf{a}$ or of its direction or of both. The difference must be obtained vectorially.

A vector may be represented as a single row of n numbers, or as a single column. The column representation is the one normally used. Such a collection of numbers written in a regular array is termed a matrix. In this specific case we have an $n \times 1$ matrix, and in general a matrix of m rows and n columns is termed an $m \times n$ matrix. A square matrix has n rows and columns, and a diagonal matrix is a square matrix which has zeros everywhere except along the diagonal. A diagonal matrix whose diagonal elements are all unity is normally taken as the unit matrix. The matrix whose elements are all zero is the zero matrix.

Matrices are not numbers, but in some cases numerical equivalents may be assigned. In a square matrix the equivalent determinant has a numerical representation and is called the "determinant" of the matrix. Here the matrix becomes a determinant simply by agreeing to assign a numerical value to it which is obtained by the standard method for evaluating a determinant. For matrices, algebraic rules may be postulated which are similar in many respects to the rules of ordinary algebra. Thus, for example, we may postulate the sum of two matrices $\|A\|$ and $\|B\|$ to be the matrix whose elements are formed from the sums of the corresponding elements of $\|A\|$ and $\|B\|$. As an example,

$$\left\| \begin{matrix} a & b \\ c & d \end{matrix} \right\| + \left\| \begin{matrix} e & f \\ g & h \end{matrix} \right\| = \left\| \begin{matrix} a+e & b+f \\ c+g & d+h \end{matrix} \right\| \tag{1}$$

We may also define the operation of multiplication. Given a matrix $\|M\|$ in which i and j are the row and column indices respectively of its elements m_{ij}, the elements of the product $\|P\| = \|M\| \, \|N\|$ are given by

$$P_{ij} = \sum_{k=1}^{n} m_{ik} n_{kj} \tag{2}$$

It is implied in this relation that the number of columns of $\|M\|$ equals the number of rows of $\|N\|$. As an illustration we may compute the product

$$\left\| \begin{matrix} a & b \\ c & d \end{matrix} \right\| \times \left\| \begin{matrix} e & f \\ g & h \end{matrix} \right\| = \left\| \begin{matrix} ae + bg & af + bh \\ ce + dg & cf + dh \end{matrix} \right\| \tag{3}$$

It will be readily established by trial that the product matrix which is obtained may depend upon the order of multiplication and $\|M\| \times \|N\|$ is not in general equal to $\|N\| \times \|M\|$. Thus the order of factors in matrix algebra must be consistently maintained. The product of a matrix $\|M\|$ and a scalar quantity S is the matrix whose elements are obtained by multiplying the elements of $\|M\|$ by S.

Matric equations for a multicompartment system.* In the notation of matrix algebra we may now rewrite eq. 1, Chapter 3:

$$\frac{d\|R\|}{dt} = \|r\| \cdot \|a\| \tag{4}$$

where
$$\|r\| = \left\| \begin{matrix} -\sum_{j \neq 1}^{n} \rho_{j1} & \rho_{12} & \rho_{13} & \cdots & \rho_{1n} \\ \rho_{21} & -\sum_{j \neq 2}^{n} \rho_{j2} & \rho_{23} & \cdots & \rho_{2n} \\ \cdots & \cdots & \cdots & \cdots & \cdots \\ \cdots & \cdots & \cdots & \cdots & \cdots \\ \rho_{n1} & \rho_{n2} & \cdots & \cdots & -\sum_{j \neq n}^{n} \rho_{jn} \end{matrix} \right\| \tag{5}$$

We represent $\|R\|$ and $\|a\|$ in matrix form since the vectors are representable as column matrices (i.e., matrices with n rows and 1 column). We may regard $\|r\|$ as a difference of two matrices and write

$$\|r\| = \|\rho\| - \|\Sigma\| \tag{6}$$

* The author was assisted in preparing this section by discussions with Mr. Don P. Engelberg. There would seem to be little point in concerning ourselves here with the alternative tensor or dyadic expressions of a linear vector transformation since, for the present purpose, they are essentially similar to the matrix representation.

The elements of the $\|\rho\|$ matrix are the ρ_{ij}, i being the row and j the column index respectively. The matrix $\|\sum\|$ is a diagonal matrix, the element of the jth row or column being the sum of all the row elements of the $\|\rho\|$ matrix. Although ρ_{ii} appears in this analysis it will cancel out in each case, and its inclusion is for symmetry purposes only.

Inspection of eq. 4 indicates that the **R** and **a** vectors do not uniquely specify the matrix $\|r\|$, which merely represents a transformation on **a** whereby it is converted into $d\mathbf{R}/dt$. In general this method of representing a linear transformation is an example of the convenience of matrix notation. It is possible to establish the $\|r\|$ matrix uniquely if we define two $n \times n$ matrices $\|R\|$ and $\|a\|$, where the rows represent the compartments but now the individual columns represent different experiments or different tracers. In this case we have

$$\|dR/dt\| = d\|R\|/dt = \|r\|\,\|a\| = \|\rho\|\,\|a\| - \|\textstyle\sum\|\,\|a\| \qquad (7)$$

as a relation between the square matrices $\|R\|$ and $\|a\|$. In order to establish $\|r\|$ as a function of the measured data, represented by $d\|R\|/dt$ and $\|a\|$, we must introduce a new matrix $\|a^{-1}\|$ defined as that matrix which, if it exists, satisfies the relation

$$\|a\|\,\|a^{-1}\| = \|1\| \qquad (8)$$

where the quantity on the right is the unit matrix. It can be shown (107) that any non-singular square matrix has an inverse whose elements are given by

$$a_{ij} = C_{ji}/|a| \qquad (9)$$

The denominator $|a|$ is the determinant of $\|a\|$ (non-zero if $\|a\|$ is non-singular) and C_{ji} is the co-factor of the element a_{ji} in the determinant of $\|a\|$. We now proceed to multiply both sides of eq. 7 by $\|a^{-1}\|$. In order to obtain the desired result the factor is applied terminally in each case (postmultiplication). Thus

$$\|dR/dt\|\,\|a^{-1}\| = \|r\|\,\|a\|\,\|a^{-1}\| - \|\textstyle\sum\|\,\|a\|\,\|a^{-1}\|$$
$$= \|r\| \qquad\qquad - \|\textstyle\sum\| \qquad (10)$$

We will be interested only in the elements of $\|r\|$ which are non-diagonal since in the present theory ρ_{ii} has no operational meaning. All the non-diagonal elements of $\|\sum\|$ are zero since it is a diagonal matrix. Equating corresponding non-diagonal elements on the two sides, we obtain

$$\rho_{jk} = \sum_i [dR_{ji}/dt][C_{ki}/\,|a|\,] \qquad (11)$$

The same relation is obtained by expanding eq. 18 in Chapter 3 by Cramer's rule if rows and columns are interchanged in the numerator and denominator.

Matric differential equations.   The matrix notation is particularly convenient in the solution of systems of first-order linear differential equations. These solutions will involve the concept of matric power series. Such series are not algebraic relations in the ordinary sense but often bear a striking similarity to the more familiar power series of elementary calculus. Thus, for example, we may write

$$e^{\|r\|} = \|1\| + \|r\| + \|r\|^2/2! + \cdots \tag{12}$$

The first term on the right is the unit matrix, the third is obtained by multiplying the matrix $\|r\|$ by itself and dividing all the elements by 2! All the additions are in the matrix sense. The reader is referred to more advanced texts for discussions of the conditions under which such series converge. From the fact that differentiation of a matrix with respect to a scalar variable can be performed within the matrix directly, it is evident that

$$\frac{d}{dt}\|t^n r\| = \frac{d}{dt}t^n\|r\| = nt^{n-1}\|r\| \tag{13}$$

if all r_{ij} are constant.

We apply this rule to the series $e^{\|tr\|}$ term by term, showing that

$$\frac{d}{dt}e^{\|tr\|} = \frac{d}{dt}e^{t\|r\|} = \|r\| + t\|r\|^2 + \frac{t^2\|r\|^3}{2!} + \cdots$$
$$= \|r\|\,e^{t\|r\|} = e^{t\|r\|}\,\|r\| \tag{14}$$

We now consider the general kinetic relations for multicompartment systems, using eq. 12, Chapter 4, written in matrix notation,

$$d\|a\|/dt = \|\sigma\|\,\|r\|\,\|a\| - \|\sigma\|\,\|\textstyle\sum\|\,\|a\| \tag{15}$$

where

$$\|\sigma\| = \begin{Vmatrix} 1/S_1 & 0 & 0 & \cdots & 0 \\ 0 & 1/S_2 & 0 & \cdots & 0 \\ 0 & 0 & 1/S_3 & \cdots & 0 \\ \cdot & \cdot & \cdot & \cdot & \cdot \\ 0 & 0 & 0 & \cdots & 1/S_n \end{Vmatrix}$$

For purposes of convenience we may write

$$d\|a\|/dt = \|\mu\|\,\|a\| \tag{16}$$

and from eq. 14 we may verify that, if $\|\mu\|$ is not a function of t, a solution may be obtained of the form

$$\|a\| = e^{(t-t_0)\|\mu\|}\,\|a_0\|$$

where $\|a_0\|$ is the original vectorial representation of the compartmental specific activities at t_0 and is not a function of t. In more sophisticated treatises the proof may be found that this is the general solution of the system of equations.

The algebraic expression of the matrix exponent may be obtained by determining the characteristic roots of the matrix $\|\mu\|$. These may be defined through the determinant Δ of the "characteristic matrix." Thus if μ_{ij} are the individual elements of $\|\mu\|$,

$$\Delta = \begin{vmatrix} (\lambda - \mu_{11}) & -\mu_{12} & \cdots & -\mu_{1n} \\ -\mu_{21} & (\lambda - \mu_{22}) & \cdots & -\mu_{2n} \\ \cdots\cdots\cdots\cdots\cdots\cdots\cdots\cdots \\ -\mu_{n1} & -\mu_{n2} & (\lambda - \mu_{nn}) \end{vmatrix} = 0 \qquad (17)$$

The n roots λ_i of this determinant relation (if separate) are the characteristic roots or "eigenvalues." There is an evident resemblance to the determinants yielding the exponential constants obtained in the earlier chapters. A number of numerical methods have been developed for their evaluation (109). In the theory outlined in this book it has always been possible to take $t_0 = 0$, i.e., to start our watches at the beginning of an experiment. Furthermore it is possible to express the matric exponential in terms of the characteristic roots in the form,

$$\|a\| = e^{\|\mu\|t} \|a_0\| = \left[\sum_{i=1}^{n} e^{\lambda_i t} \|X_i\| \right] \|a_0\| \qquad (18)$$

where the $\|X_i\|$ are constant square matrices defined by

$$\|X_i\| = \frac{1}{\prod\limits_{j \neq i} (\lambda_j - \lambda_i)} \prod_{j \neq i} [\lambda_j \|1\| - \|\mu\|] \qquad (19)$$

We thus have a general solution of the multicompartment system provided the $\|\mu\|$ matrix does not vary with time. Generally this restriction will mean that the system is in a steady or quasi-steady state. It is possible to extend the matrix method to include the situation where $\|\mu\|$ is a function of time (55, 107). Here for $t_0 = 0$ we have

$$\|a\| = \left\{ \|1\| + \int_0^t \|\mu\| \, ds + \int_0^t \|\mu\| \, ds \int_0^s \|\mu\| \, ds_1 + \cdots \right\} \|a_0\|$$

$$= \Omega_0{}^t \|a_0\| \qquad (20)$$

where $\Omega_0{}^t$ is called the "matrizant" of $\|\mu\|$. The difficulty is that the matrizant will often be too cumbersome for practical evaluation.

It is clear from the previous discussion that matrix methods have a certain advantage in expressing the complex relations of tracer kinetic theory. The methods are mathematically elegant and economical. Where general relations are to be expressed and discussed this system of representation is to be preferred. Actually, for most individual practical cases in systems of a few compartments it would seem debatable whether it is necessary for those untrained in matrix calculus to master the necessary elements required for tracer kinetics. Preference will undoubtedly depend upon the individual inclinations of the research worker. Whatever the decision may be, upon following the theory through for individual cases it will soon become evident that matrix methods do not shorten the labor of reducing the solutions of the differential equations to numerical results. Whether we prefer to evaluate a secular determinant of the form of eq. 19 in Chapter 4, or whether we determine the eigenvalues in eq. 17 here, there is little difference in the ultimate labor required, and the computation is essentially the same.

Constant Linear Systems

Matric models for compartmental systems. The advantage of using matrices in obtaining general conclusions concerning constant linear systems has been recognized for many years in the field of electrical engineering. The postulation of electric analogs for tracer systems implies the equivalence between the mathematical methods of linear electrical systems and linear tracer systems (110). The analysis will be applied only to constant systems in which the matrix $\|\mu\|$ is not a function of time. In this case, in principle at least, the requirements of multiple tracers or separate experiments do not place a serious restriction on the tracer technique. A single experiment conducted over an extended period of time may be considered as a continuous series of experiments under identical conditions since the constant properties of the system are invariable. The matrix method can be used as indicated either to predict the behavior of a given model or to proceed in reverse and try to solve for the $\|\mu\|$ matrix from given data.

The rates of change of a in various compartments could be taken at different times and the necessary information obtained to solve the system. It is preferable, however, to attempt to obtain a multiple exponential fit to the data. If the fitted relation is of the form

$$a_i(t) = \sum_j X_{ij} e^{-\lambda_j t} \tag{21}$$

then the X_{ij} may be regarded as the elements of a matrix $\|X\|$ where i and j are row and column indices respectively. (This matrix differs from the matrices $\|X_i\|$ in eq. 18.) We insert this relation into eq. 16. Upon equating to zero the coefficients of like exponentials (110), we obtain a matrix relation

$$\|\mu\| \, \|X\| = \|X\| \, \|-\lambda\| \tag{22}$$

where $\|-\lambda\|$ is a diagonal matrix of the characteristic roots $-\lambda_i$. Postmultiplying both sides by $\|X^{-1}\|$, we may uniquely determine the elements of the $\|\mu\|$ matrix and thus obtain the transport rates, i.e.,

$$\|\mu\| = \|X\| \, \|\lambda\| \, \|X^{-1}\| \tag{23}$$

In so doing it is assumed that there are no degeneracies due to possible confluence of one or more characteristic roots, and that the matrix elements X_{ij} and the characteristic exponents λ_i are all precisely determinable.

Permissible models for underdetermined systems. In many instances, particularly in biological research, it may not be possible to obtain all the elements of the matrices. It is also worthwhile in any general analysis to consider the effect of possible constraints in the system which might occur, say, if certain transport rates were known to be zero. Where certain elements of the $\|X\|$ matrix or $\boldsymbol{\lambda}$ vector are missing, we may regard the system as underdetermined. To illustrate this concept, if we have two equations in x and y both variables are uniquely determined. However, if we have only one, the system is underdetermined and only a single relation between them is imposed. Within the restriction of this relation they are both free to vary. In a more general situation, eq. 23 no longer uniquely determines the $\|\mu\|$ matrix elements and certain relations among them are permitted. Berman and Schoenfeld (110) formulate this problem by defining a non-singular matrix $\|P\|$ which represents a similarity transformation on the matrix $\|\mu\|$. In this way they are able to describe all possible alternate matrices $\|\mu'\|$ which are compatible with the available experimental data, yielding

$$\|\mu'\| = \|P\| \, \|\mu\| \, \|P^{-1}\| \tag{24}$$

The actual methods for obtaining the transformations may be obtained from their original paper.

Of course if a certain number of elements of $\|\mu\|$ are not required, the matrix may be expressible with a rank lower than n. It may then be possible to determine lower rank $\|X\|$ and $\|\lambda\|$ relations which will permit the unique determination of the reduced $\|\mu\|$. Situations might also arise where the system is overdetermined. Here all the additional restrictions might be compatible, but because of experimental or truncation error it is more likely that they will not. In such a case methods exist for fitting the system by a compromise model, using the principle of least squares or other techniques (111). Normally it would seem unlikely that this situation would exist in practice. Usually it will already be rather difficult to obtain all the required eigenvalues from experimental data, let alone extra eigenvalues.

Statistics and Curve Fitting

The information which we can obtain by the tracer method concerning the basic kinetic properties of a compartmental system will, of course, be no better than the original data. It is a particularly unfortunate fact that tracer measurements in many physiological systems are subject to a considerable amount of fluctuation. This "noise" may often seriously obscure the informational content of the measurements. If the tracer method is to succeed at all it becomes necessary to apply all our resources to this information retrieval problem. We are thus concerned in this section with the problem of making the "best fit" to data, particularly in the case of specific activity curves obtained in compartments of a kinetic system.

Estimation of "best values." The underlying philosophy of estimating best values is that, if a series of repeated measurements of a quantity are made, their result can be combined to obtain a better value than any of the individual measurements. It is also inherently assumed that it is less economical under the particular circumstances to attempt to develop a more precise technique than to make the required series of repeated determinations by existing procedures. Perhaps the system under study is inherently variable and cannot be improved in this regard. In general there is no accepted criterion among statisticians for a "best value," though in particular instances good criteria may suggest themselves (112). The decision in a given case may be considered as part of the "art" rather than the science of statistics. The most widely accepted basis for estimation is the "likelihood" function, and the best value of a parameter is the one which maximizes this function.

In obtaining this estimate the procedure will generally depend on how the data are distributed when the frequency with which values fall in a given class interval is plotted as an ordinate against class interval as abscissa. In many instances it happens that the individually measured values cluster fairly closely together in a more or less symmetrical distribution with a well-defined peak. If it happens that the measurements are influenced by a large number of small effects which act together additively to influence the results we may then recognize that the central-limit theorem of probability should apply, yielding a normal distribution of results. If this justification cannot be made it may become necessary to assume intuitively that a normal distribution holds or that the results of violation of normality will not be serious for data which cluster about some peak value.

Often we shall represent values as ordinates rather than as frequencies about some peak value. Under these conditions, given a series of values y_i determined for the x_i, we estimate the best value $\hat{y}$ as the value for which the sum of the squares of the deviations of individual measurements from $\hat{y}$ is a minimum. This "least squares" fit is often quite adequate for estimation purposes. For independent normally distributed quantities the result yields the mean $\bar{y}$ as the least-squares estimate.

We may be able to improve the estimation if we decide that certain measurements are better than others and assign weights to them. Thus if one of the y_i were the mean of two separate measurements, it should receive a weight of 2. If a measurement is made with an instrument which has a constant percentage error, the variability of large values would be greater than of small, and the latter would receive more weight. The concept of constant percentage error is often employed by physicists for expressing precision of results. This is feasible only for small errors since large fluctuations have a skewed distribution. It is of course necessary in assigning weights to consider objectively all sources of error. Often there will be a dilution error which may require consideration, and other sources of fluctuation include background errors which apply more to small values than to large. Ideally the experiment would be designed to minimize the subjective feeling of the worker as to the precision of the result. Nevertheless it may be dangerous to rule this out entirely, since the instinct of a trained laboratory worker is a valuable commodity, not to be underrated.

There are, of course, instances when the data are not entirely normally distributed. Occasionally it may happen that some measurement differs from the rest by a large amount. If there are several of these discrepant readings, particularly if they all deviate in the same direction, we may suspect that something has occasionally gone wrong with the method. If other evidence can also be advanced to justify this suspicion it may be legitimate to reject these discordant results, particularly after performing a significance test. Of course even normally distributed data can occasionally yield a highly discordant value, and rejecting this or similar fluctuant results, although not influencing the mean very much, may yield an underestimate of the uncertainty of the determinations.

Regression and curve fitting. As further determinations of y_i are made, it may appear to the investigator that the values are drifting progressively with time. Consider, for example, a series of sequential measurements of the activity of a radioactive isotope. If we plot the results as a function of time and if the isotope has a long half life, the best fit may be $y = \bar{y}$, where $\bar{y}$ is the arithmetic mean of the individual y values. Nevertheless it is known, from independent information, that decay is occurring, and an analysis may therefore be made in search of a negative trend in the data. We now make a linear regression analysis by fitting the relation $\hat{y} = -mt + b$ to the data. Here, as in many other instances, time can be measured with high precision, and thus no fluctuation in its values need be assumed. We thus are permitted in this case to consider only a regression of $\hat{y}$ on t, which greatly simplifies the labor. Using the least-squares estimates of the parameters m and b, we may be able to test for a significant difference of m from zero. This test does *not* tell us that there is or is not a trend in the data. Instead we obtain only an estimate of the chances that we are wrong if we say that m does not differ significantly from zero. This estimate in turn is not an absolute *measure* of the odds but may be an unbiased *estimate* of them.

After the preliminary linear analysis we may not be satisfied that a linear fit is the best prediction. In search of the quadratic term in the power series expansion of $\hat{y}(t)$ near $y = 0$, we attempt to fit a parabola to the data. We cannot definitely determine whether the fit is better, but we may choose the curve yielding the lower sum of squares as the best one. Given enough determinations over a large enough period of time with small enough fluctuation, suppose that we conclude from the sum of squares that a parabola fits the data better than a linear relation. We must not conclude then that the least-squares parabola is the *correct* function $\hat{y}(t)$, and we may, in fact, obtain an absurd result in which $\hat{y}(t)$ initially increases and reaches a maximum before declining. In this event we should find that alternative fits are almost as good and that they do not differ significantly from the one we have obtained. We might now decide that the activity must always decline from the initial value and insist that the parabola which we fit to the data must at most have zero derivative at $t = 0$. This would permit a better fit to the data based on our intuitive concept of the system.

Although curve fitting by statistical procedures can furnish an estimate of the best fit, an equally important aim is to obtain some idea of the variability of the measurements. Many of the commonly employed methods assume that this variability is uniform from one subgroup of the data to another. If not, it becomes necessary to apply weights to some of the measurements to homogenize the variability. These weights are now introduced not because any objective knowledge is available concerning the precision of the values but because greater fluctuation over one part of a curve suggests less precision. Here, in the case of y_i which vary quasi-exponentially with time it may happen that the fractional error tends to be constant. If the data are now transformed to a log scale, this tends to make the *absolute* fluctuation uniform. If at the same time it linearizes the data and reduces the labor of curve fitting, this type of transformation may prove attractive. If the fluctuation in the data is not great, the distribution of the y_i about the regression line may still be reasonably normal and a least-squares fit may still be acceptable. Of course the best fit will be to the *log* of the data, not to the data themselves. If the data are not too fluctuant this distinction may not be very serious. If there is greater variability the bias introduced into $\hat{y}$ by estimating the best *log $\hat{y}$* may prove embarrassing.

From these few illustrations of the misconceptions and hazards of regression analysis of data it will become evident that the best results will be achieved after consultation with an experienced statistician before the data are obtained. We should not expect to obtain good results from poor data, and statistical analysis should be optimally combined with common sense to be maximally effective. Least-squares fitting provides maximum likelihood only under certain conditions, and even maximum likelihood estimates cannot be said to be the best ones, though we may often prefer them.

Fitting of multiple exponentials to data: general considerations.
Let us consider the problem of fitting a series of exponentials of the form

$$\hat{y} = \sum_n A_n e^{-\lambda_n x} \tag{25}$$

to a set of data whose coordinates are y_i, x_i (Fig. 87). The problem is to determine the best values of the parameters λ_n and A_n which will minimize the sums of the squares of the errors $\hat{y} - y_i$. A relatively large number of tracer experimenters have used the method of exponential "peeling" for estimating these parameters (28). The method has the advantage that it can be relatively easily employed. A certain amount of subjective common sense judgment is also permitted in the method. It has a number of disadvantages, however. Since it involves curve fitting on semi-log coordinates it may tend to bias the estimates. An arithmetic mean on logarithmic coordinates becomes a geometric mean in rectangular representation. Like all graphic methods, it is not readily adaptable to objective estimation of error. Subjective elements of personal bias may also creep into the choice of the best fit. Since the estimates of all the more rapidly varying exponential terms depend upon the estimates of the slower components, errors in estimation may readily cumulate. It will frequently provide an excellent method for preliminary analysis of data, and in the case of iterative methods for curve fitting it may assist in obtaining good initial guesses of the parameters. These may then be improved iteratively.

The principal difficulty in exponential curve fitting is the insensitivity of the data to the number of possible exponentials which might be present.

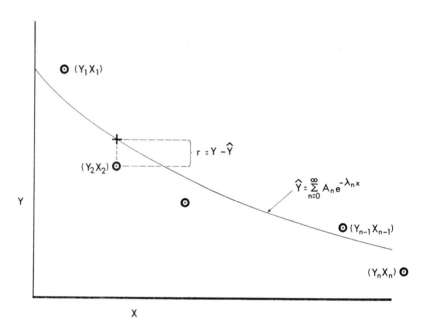

Fig. 87. Fitting of exponential sums to data.

This can be readily appreciated, for example, if a series of three exponentials is synthesized and an attempt is made to decompose the curve once more back into its components. It will often be noted that two exponential terms are practically as good as three unless the data are of extremely high accuracy. It would thus appear that the biggest problem in exponential curve fitting is the determination of the number of exponential components required. Here the experimenter will be greatly aided if any advance information is available concerning the expected number of components. Standard methods for fitting curves to data by least squares require advance knowledge of the number of parameters to be fitted before the computations begin. Lacking this knowledge, we should have to begin with some lower number and repeat all the computations successively with larger numbers until significance tests did not indicate further improvement in the fit. If we lacked advance information and used data from an incomplete experiment, it would be quite easy for the result to yield an incorrect number purely on the basis of chance.

Non-linear least squares. If we assume that a properly designed experiment has been conducted it may be possible to obtain a satisfactory least-squares fit of eq. 25 to data in some cases at least. If the parameters were linearly related to $\hat{y}$ it would be possible in a single computation to solve the classical normal equations. In the present instance, the λ's do not appear linearly and it is necessary to use the method of "linearization" (113, 114). In this method the relation is approximated by the first two terms of a power series using initially "guessed" values of the A_n and λ_n. A set of pseudo-normal equations is then solved, and from them corrections to the parameters are obtained. After correction, the new values are then inserted into the relation and further iterations computed until the amount of correction becomes negligible.

Such methods have been described by Deming (114) not only for curve fitting but also for other statistical adjustment problems. Here we will consider that all the variation is in y since time may be precisely measured. We will describe the method as it might apply to a three-exponential fit, and the generalization to a larger number will become evident from the simpler system.

The first step in the procedure is to obtain the negative partial derivatives of the right-hand side of eq. 25 with respect to the parameters A_n and λ_n. These are

$$-\partial/dA_1 = -e^{-\lambda_1 x}, \qquad -\partial/dA_2 = -e^{-\lambda_2 x}, \qquad -\partial/dA_3 = -e^{-\lambda_3 x},$$
$$-\partial/d\lambda_1 = xA_1 e^{-\lambda_1 x}, \qquad -\partial/d\lambda_2 = xA_2 e^{-\lambda_2 x}, \qquad -\partial/d\lambda_3 = xA_3 e^{-\lambda_3 x}$$
$$(26)$$

We now define the initial guesses of the A_n and λ_n as A_{n0} and λ_{n0}. For each data point, using these guessed parameters, we compute the arithmetic values of the derivatives; thus for the ith data point we have

$$^i n_1 = -e^{-\lambda_{10} x_i}, \qquad ^i n_2 = -e^{-\lambda_{20} x_i}, \qquad ^i n_3 = -e^{-\lambda_{30} x_i}$$
$$^i n_4 = x_i A_{10} e^{-\lambda_{10} x_i} \qquad ^i n_5 = x_i A_{20} e^{-\lambda_{20} x_i} \qquad ^i n_6 = x_i A_{30} e^{-\lambda_{30} x_i} \qquad (27)$$

We now obtain all possible cross products and sum them over all the data points x_i.

Thus, using the notation of Gaussian brackets, we define

$$[n_{11}] = \sum_i {}^i n_1 {}^i n_1$$
$$[n_{12}] = \sum_i {}^i n_1 {}^i n_2, \text{ etc.} \qquad (28)$$

We also numerically evaluate

$$^{i}n_0 = y_i - \hat{y}(x_i, A_{n0}, \lambda_{n0}) = y_i - \sum_n A_{n0}e^{-\lambda_{n0}x_i}$$

at every data point, and define the cross product sums

$$[n_{10}] = \sum_i {}^{i}n_1{}^{i}n_0,$$

$$[n_{20}] = \sum_i {}^{i}n_2{}^{i}n_0, \text{ etc.} \tag{29}$$

From these quantities we form the pseudo-normal equations

$$[n_{11}]u_1 + [n_{12}]u_2 \cdots [n_{16}]u_6 = [n_{10}]$$
$$[n_{21}]u_1 + [n_{22}]u_2 \cdots [n_{26}]u_6 = [n_{20}]$$
$$\cdots\cdots\cdots\cdots\cdots\cdots\cdots\cdots\cdots\cdots\cdots$$
$$[n_{61}]u_1 + [n_{62}]u_2 \cdots [n_{66}]u_6 = [n_{60}] \tag{30}$$

The solution by matrix inversion is particularly convenient since the $\|n\|$ matrix is symmetrical.* Upon obtaining the u's, the improved values of the parameters are obtained as

$$A_{11} = A_{10} - u_1, \quad A_{21} = A_{20} - u_2, \quad A_{31} = A_{30} - u_3$$
$$\lambda_{11} = \lambda_{10} - u_4, \quad \lambda_{21} = \lambda_{21} - u_5, \quad \lambda_{31} = \lambda_{30} - u_6 \tag{31}$$

This method has the advantage in principle that it can obtain a weighted least-squares fit to data with any spacing of x_i values. Since a large number of data points must be used in order to obtain satisfactory estimates of the "best values," the arithmetic labor is prodigious, and few will attempt its use except through the agency of a large digital computer.† One weakness of iterative methods of this type is the uncertain convergence. With good data and the best initial guess this problem may be less critical, but the phenomenon of "overshoot" has proved serious in some instances. The incorporation into the computer program of some method of arbitrarily applying an adjustment factor to the u values when they appear to over-correct has been suggested. A discussion of the practical use of iterative methods for exponential curve fitting by machine methods has been given by Worsley et al. (115).

---

* It must be kept in mind that inversion of large matrices can be subject to serious error unless sufficiently precise computations are performed (109, 111).

† A description of a recent program of another type for fitting of model systems to tracer data and written for the IBM 704 computer may be obtained from Dr. M. Berman, Institute of Arthritis and Metabolic Diseases, National Institutes of Health, Bethesda, Maryland.

Uniform spacing of abscissal values. Because, under certain conditions, expressions of the form of eq. 25 satisfy linear difference equations with constant coefficients, methods of interpolation and of curve fitting have been developed for $\hat{y}$ which require uniform spacing and other special relations on the distribution and number of values of x_i. The classical example of this approach is Prony's method (113). Householder (116) has discussed the method critically and describes an iterative procedure whereby it can be made to yield a valid weighted least-squares fit. Included also is a method whereby the extension to a greater number of exponentials in the fit can be achieved, utilizing the material of the previous computations. In a recent Ph.D. thesis R. G. Cornell has described a non-iterative method including a comprehensive discussion of the sufficiency consistency, bias, and efficiency of the estimators employed in the method (117). Difficulties have been encountered in the direct application of Prony's method (118). Important words of wisdom for those who have difficulties with curve fitting will be found in a recent paper by Cornfield et al. (119).

A semi-objective "peeling" technique. One objection which may be registered to all the curve fitting procedures which we have discussed up to this point is that they apply to only one set of data. If the data from more than one compartment are analyzed it may be found that the least-squares fit for one specific activity curve yields one set of exponential constants but a different set may be obtained in other compartments. In some instances this may be a chance result due to the imperfection of the analysis. If the exponential constants from all compartments are required to be alike, the fit in each compartment may be almost as good and, at the same time, may now be consistent with the unified model of the system whose eigenvalues must be the same in any part of it.

A procedure which is an objective modification of the "peeling" technique suggests itself. We plot the terminal portions of the separate curves semi-logarithmically to estimate a preliminary rate constant λ_{10}. This estimation need not be precise, and if the experimental system is a closed one the first estimate $\lambda_{10} = 0$ may suffice. We now proceed to multiply the y_i for the largest x_i by $e^{\lambda_{10} x}$. If the data were precise and if the estimate were sufficiently good, the resulting product values would become constant in time for large x. Actually this will probably not occur so a linear regression curve may now be passed through the data points. If parabolic regression gives a better fit we may adopt the slope of the terminal portion of the parabola as the "best value" of the correction to be applied to λ_{10} to obtain the exponential constant λ_1. For small slope the linear approximation to the exponential will be valid. We now perform this estimation for each compartment, obtaining in each case a best value of λ_1. If these do not differ significantly a simple average may be sufficient. If they appear to differ it may be possible to obtain a weighted average which will still suffice, although this should now warn us that the simple compartmental model we have chosen is perhaps unsatisfactory.

Given the best value of λ_1 we now multiply the terminal y_i by $e^{\lambda_1 t}$ and form a sequential set of means starting with the largest x_i and proceeding inward. At the same time we sequentially compute the standard error and compute the Student's t value. As soon as we encounter a y_i value which significantly exceeds the mean at the 5 per cent level we terminate the procedure. The resulting mean will yield the "best" value of A_1 for each compartment. (In compartments where the data contain a maximum or do not decline monotonically it may be necessary to terminate when a_{yi} significantly falls short of the mean.) Upon obtaining the value of A_1 for each compartment, we then proceed to subtract $A_1 e^{-\lambda_1 t}$ from all the data for that compartment and proceed as before.

Concluding remarks concerning curve fitting. Most of the few discussions in the literature concerning the fitting to tracer data of multiple exponentials consider compartments separately and independently. In the preceding section we have considered how the data may be combined to satisfy the requirement that the eigenvalues be the same for every compartment. Other requirements also suggest themselves. In compartments which represent a precursor relationship to one or more neighbors it will be necessary to require the specific activity curve of the precursor to intersect the curve of the subsidiary at the maximum of the latter. Other restrictions include the requirement that the specific activity of any initially unlabeled compartment be truly zero. The fact that the zero intercept of the least-squares fitted curve does not differ significantly from zero will not be considered sufficient for most workers. Other requirements include the restriction of the specific activity for all compartments to positive values and the imposition of a zero initial slope for the specific activity curve in all compartments which do not adjoin an initially labeled one. Actually Deming's procedures may be extended to include auxiliary constraints. No example of such extensions has been encountered in the literature of tracer kinetics at this time.

Appendix

First Traversals of the Random Walk

If the randomizing properties of a labyrinth change discontinuously (for example, if there is a sudden change in grain size in a bed of glass beads), there will be an effect on the stochastic distribution of traversal times. We will consider only the effect of complete removal of randomizing elements beyond some boundary in the system. In this case if a particle of label passed this boundary it would immediately proceed on its way and not walk randomly back into the labyrinth. Particles appearing at the outflow of the labyrinth will only be those which have experienced a single traversal of the boundary. As in eq. 33 in Chapter 9, in a continuous column without a boundary, we have the fraction of label between x and $x + dx$ as

$$F(x) = (\kappa^2 \pi \tau)^{-\frac{1}{2}} e^{-(x-\tau)^2/\kappa^2 \tau}$$

Equation 33 holds only at $x = 1$; thus the present relation is a more general one. Introducing a boundary, we may imagine the distribution to be a combination of particles of label which are passing the boundary for the first time plus those making second, third, etc., traversals in either direction (120). The fraction of label particles which have crossed the boundary at $x = 1$ can be considered as particles which have crossed forward at least once. They are thus "crossed" particles, which we shall represent as $F(x \geq 1)$.

We now must consider the problem of determining the fraction of particles which, having crossed, have again returned across the boundary and are now in the region $x < 1$. This problem would be rather complex but may be simplified by the method of images. Because we are concerned with a type of diffusion problem, we conclude that the density of particles cannot be discontinuous in the absence of sources or sinks. If at any moment such a jump in density were to occur it would immediately disappear owing to the random motions of diffusion. For this reason the density of crossed particles at the boundary will be continuous. This means that whatever the form of $F(x < 1)$, it must equal $F(x \geq 1)$ at the boundary. Furthermore, since all particles are subject to the processes of diffusion with drift the two functions must have the same form except that they can have different normalizing factors and can be represented in a shifted coordinate system. From these considerations it is inferred that $F(x < 1)$ is of the form

$$\frac{e^{4/\kappa^2}}{\sqrt{\kappa^2 \pi \tau}} \left\{ \exp \left[\frac{-(x - 2 - \tau)^2}{\kappa^2 \tau} \right] \right\}$$

This relation behaves as though an initial delta function of label had been initially introduced at $t = 0$ into the system at the "image point" located at $x = 2$, symmetrically placed with respect to $x = 0$ and the boundary. We now determine the total fraction F_c of "crossed" label expressed as a function of τ by integrating over the *space* coordinates,

$$F_c = \int_{-\infty}^{1} F(x < 1) \, dx + \int_{1}^{\infty} F(x \geq 1) \, dx$$

$$= (\pi)^{-\frac{1}{2}} \left\{ e^{4/\kappa^2} \int_{-\infty}^{\alpha} e^{-y^2} \, dy + \int_{\beta}^{\infty} e^{-y^2} \, dy \right\}$$

where $\alpha = (1 + \tau)/\kappa\sqrt{\tau}$ and $\beta = (1 - \tau)/\kappa\sqrt{\tau}$.

Since the rate of increase in crossed particles with respect to τ is $dF_c/d\tau$, which is the rate of first crossings, $F(\tau)$, we have

$$\frac{dF_c}{d\tau} = F(\tau) = (\pi)^{-\frac{1}{2}} \left\{ e^{4/\kappa^2 - \alpha^2} \frac{d\alpha}{d\tau} - e^{-\beta^2} \frac{d\beta}{d\tau} \right\}$$

$$= \frac{e^{-(1-\tau)^2/\kappa^2\tau}}{\kappa\sqrt{\pi \tau}^{\frac{3}{2}}}$$

Some Properties of the Random-Walk Function

The Laplace transform $= \exp\left[(2/\kappa^2)(1 - \sqrt{1 + \kappa^2 s})\right]$.

If we define the ith moment about the origin as

$$m_i = \int_{0}^{\infty} \tau^i \frac{\exp\left[\dfrac{-(1 - \tau)^2}{\kappa^2\tau} \right]}{\sqrt{\pi\kappa\tau}^{\frac{3}{2}}} \, d\tau$$

We have

$$m_0 = \int_0^\infty \frac{(\tau)^0 \exp\left[\dfrac{-(1-\tau)^2}{\kappa^2\tau}\right]}{\sqrt{\pi\kappa}\tau^{3/2}}\, d\tau = \frac{e^{2/\kappa^2}}{\sqrt{\pi\kappa}} \int_0^\infty \frac{\exp\left[\dfrac{-1}{\kappa^2\tau} - \dfrac{\tau}{\kappa^2}\right]}{\tau^{3/2}}\, d\tau$$

We substitute $\tau = 1/\kappa^2 x^2$, obtaining

$$m_0 = \frac{2e^{2/\kappa^2}}{\sqrt{\pi}} \int_0^\infty \exp\left[-x^2 - \left(\frac{1}{\kappa^4 x^2}\right)\right] dx$$

From standard integral tables (e.g., Peirce 495) this becomes

$$m_0 = \exp\left[\frac{2}{\kappa^2} - \frac{2}{\kappa^2}\right] = 1$$

Similarly

$$m_1 = \int_0^\infty \frac{\exp\left[\dfrac{-(1-\tau)^2}{\kappa^2\tau}\right]}{\sqrt{\pi}\tau^{1/2}}\, d\tau = 1$$

Here we substitute $\tau = \kappa^2 x^2$.

From these relations we may compute all higher moments thus,

$$m_i = \frac{e^{2/\kappa^2}}{\sqrt{\pi\kappa}} \int_0^\infty \tau^{i-3/2} \exp\left[\frac{-1}{(\kappa^2\tau)} - \frac{\tau}{\kappa^2}\right] d\tau$$

integrating by parts $u = \tau^{i-3/2} e^{-1/\kappa^2\tau}$, $dv = e^{-\tau/\kappa^2}\, d\tau$ thus

$$m_i = \int_0^\infty \frac{\exp\left[\dfrac{-(1-\tau)^2}{\kappa^2\tau}\right]}{\sqrt{\pi\kappa}\tau^{3/2}} \left[\kappa^2\left(i - \frac{3}{2}\right)\tau^{i-1} + \tau^{i-2}\right] d\tau$$

$$= \kappa^2\left(i - \frac{3}{2}\right)m_{i-1} + m_{i-2}$$

Thus we have

$$m_0 = 1$$
$$m_1 = 1$$
$$m_2 = 1 + \kappa^2/2$$
$$m_3 = 1 + 3\kappa^2/2 + 3\kappa^4/4$$
$$m_4 = 1 + 3\kappa^2 + 15\kappa^4/4 + 15\kappa^6/8$$

Principal landmarks.   The maximum value is located at the positive root of $\tau^2 + 3\kappa^2\tau/2 - 1 = 0$. The inflection points are at the positive real roots of

$$\tau^4 + 3\kappa^2\tau^3 + [(15\kappa^4/4) - 2]\tau^2 - 5\kappa^2\tau + 1 = 0.$$

A Table of the Random-Walk Function

τ	Kappa, 0.10 $F(\tau)$	τ	Kappa, 0.20 $F(\tau)$
0.50		0.50	0.00002973
0.55		0.55	0.00069561
0.60		0.60	0.00772452
0.65	0.00000007	0.65	0.04839826
0.70	0.00002511	0.70	0.19355371
0.75	0.00208791	0.75	0.54078306
0.80	0.05312728	0.80	1.12951490
0.85	0.51014504	0.85	1.85723100
0.90	2.17526340	0.90	2.50261430
0.95	4.68329630	0.95	2.85257880
1.00	5.64189570	1.00	2.82094780
1.05	4.13273600	1.05	2.47036000
1.10	1.97025700	1.10	1.94806190
1.15	0.64665361	1.15	1.40256070
1.20	0.15311047	1.20	0.93263376
1.25	0.02720118	1.25	0.57831169
1.30	0.00374848	1.30	0.33714483
1.35	0.00041219	1.35	0.18607552
1.40	0.00003705	1.40	0.09780515
1.45	0.00000278	1.45	0.04921033
1.50	0.00000017	1.50	0.02380663
1.55		1.55	0.01111631
1.60		1.60	0.00502699
1.65		1.65	0.00220813
1.70		1.70	0.00094459

τ	Kappa, 0.30 $F(\tau)$	τ	Kappa, 0.40 $F(\tau)$
0.50	0.02056372	0.50	0.17528298
0.55	0.07710819	0.55	0.34629811
0.60	0.20906366	0.60	0.57321036
0.65	0.44209303	0.65	0.82879316
0.70	0.76954863	0.70	1.07827950
0.75	1.14706340	0.75	1.28996560
0.80	1.50797480	0.80	1.44215860
0.85	1.78830560	0.85	1.52540960
0.90	1.94680930	0.90	1.54113950
0.95	1.97251240	0.95	1.49843070
1.00	1.88063190	1.00	1.41047390
1.05	1.70227840	1.05	1.29157130
1.10	1.47348820	1.10	1.15504850
1.15	1.22700180	1.15	1.01207250

Kappa, 0.30 (*cont.*)		Kappa 0.40 (*cont.*)	
τ	$F(\tau)$	τ	$F(\tau)$
1.20	0.98782980	1.20	0.87119460
1.25	0.77208315	1.25	0.73838524
1.30	0.58791707	1.30	0.61735380
1.35	0.43745581	1.35	0.50999169
1.40	0.31887959	1.40	0.41683356
1.45	0.22821626	1.45	0.33747481
1.50	0.16066374	1.50	0.27091813
1.55	0.11144396	1.55	0.21584066
1.60	0.07627593	1.60	0.17078840
1.65	0.05157733	1.65	0.13430847
1.70	0.03449472	1.70	0.10503231

Kappa, 0.50		Kappa, 0.60	
τ	$F(\tau)$	τ	$F(\tau)$
0.25	0.00111402	0.15	0.00002502
0.30	0.00998579	0.20	0.00144988
0.35	0.04358439	0.25	0.01452190
0.40	0.12187224	0.30	0.06126069
0.45	0.25401928	0.35	0.15882053
0.50	0.43192778	0.40	0.30510380
0.55	0.63432762	0.45	0.48139769
0.60	0.83556589	0.50	0.66318098
0.65	1.01319660	0.55	0.82901997
0.70	1.15201080	0.60	0.96460018
0.75	1.24479640	0.65	1.06303790
0.80	1.29110380	0.70	1.12336680
0.85	1.29521700	0.75	1.14855230
0.90	1.26412290	0.80	1.14372110
0.95	1.20586340	0.85	1.11483860
1.00	1.12837910	0.90	1.06783920
1.05	1.03880720	0.95	1.00812350
1.10	0.94313423	1.00	0.94031598
1.15	0.84609674	1.10	0.79472660
1.20	0.75123746	1.20	0.65206288
1.25	0.66104512	1.30	0.52340828
1.30	0.57713024	1.40	0.41324776
1.35	0.50040488	1.50	0.32216227
1.40	0.43124928	1.60	0.24869083
1.45	0.36965549	1.70	0.19049387
1.50	0.31534686	1.80	0.14502145
1.55	0.26787217	1.90	0.10986196
1.60	0.22667832	2.00	0.08289761

Kappa, 0.50 (*cont.*)		Kappa, 0.60 (*cont.*)	
τ	$F(\tau)$	τ	$F(\tau)$
1.65	0.19116407	2.10	0.06235107
1.70	0.16071842	2.20	0.04677497
1.75	0.13474707	2.30	0.03501545
1.80	0.11268938	2.40	0.02616687
1.85	0.09402839	2.50	0.01952664
1.90	0.07829575	2.60	0.01455458
1.95	0.06507310	2.70	0.01083831
2.00	0.05399095	2.80	0.00806474

Kappa, 0.70		Kappa, 0.80	
τ	$F(\tau)$	τ	$F(\tau)$
0.15	0.00074662	0.15	0.00654153
0.20	0.01313933	0.20	0.05312729
0.25	0.06534430	0.25	0.16772915
0.30	0.17498341	0.30	0.33442403
0.35	0.33137197	0.35	0.51651603
0.40	0.50763696	0.40	0.68315369
0.45	0.67718518	0.45	0.81725262
0.50	0.82170212	0.50	0.91324545
0.55	0.93208991	0.55	0.97263149
0.60	1.00635080	0.60	1.00035040
0.65	1.04693720	0.65	1.00248490
0.70	1.05858360	0.70	0.98501163
0.75	1.04682840	0.75	0.95322277
0.80	1.01712950	0.80	0.91152972
0.85	0.97440014	0.85	0.86346378
0.90	0.92281637	0.90	0.81176709
0.95	0.86578332	0.95	0.75851454
1.00	0.80598512	1.00	0.70523697
1.10	0.68577330	1.10	0.60266657
1.20	0.57281094	1.20	0.50926485
1.30	0.47211921	1.30	0.42701295
1.40	0.38533796	1.40	0.35611513
1.50	0.31222733	1.50	0.29586981
1.60	0.25160895	1.60	0.24517344
1.70	0.20192323	1.70	0.20280072
1.80	0.16154144	1.80	0.16755276
1.90	0.12892776	1.90	0.13832896
2.00	0.10271275	2.00	0.11415568
2.10	0.08171686	2.10	0.09419152
2.20	0.06494703	2.20	0.07772061
2.30	0.05158058	2.30	0.06414026

Kappa, 0.70 (cont.)		Kappa, 0.80 (cont.)	
τ	$F(\tau)$	τ	$F(\tau)$
2.40	0.04094359	2.40	0.05294687
2.50	0.03248876	2.50	0.04372184
2.60	0.02577432	2.60	0.03611849
2.70	0.02044540	2.70	0.02985052
2.80	0.01621797	2.80	0.02468187
2.90	0.01286531	2.90	0.02041815
3.00	0.01020684	3.00	0.01689946
3.10	0.00809896	3.10	0.01399427
3.20	0.00642762	3.20	0.01159444
		3.30	0.00961106
		3.40	0.00797101
		3.50	0.00661414
		3.60	0.00549097

Kappa, 0.90		Kappa, 1.00	
τ	$F(\tau)$	τ	$F(\tau)$
0.05	0.00000001	0.05	0.00000073
0.10	0.00089999	0.10	0.00541551
0.15	0.02821711	0.15	0.07860251
0.20	0.13486718	0.20	0.25712111
0.25	0.31181643	0.25	0.47572121
0.30	0.50788200	0.30	0.67049497
0.35	0.68210541	0.35	0.81483091
0.40	0.81572374	0.40	0.90671337
0.45	0.90560973	0.45	0.95425383
0.50	0.95641096	0.50	0.96788291
0.55	0.97551042	0.55	0.95715370
0.60	0.97046093	0.60	0.92979308
0.65	0.94790720	0.65	0.89167627
0.70	0.91327141	0.70	0.84711065
0.75	0.87078343	0.75	0.79917510
0.80	0.82364347	0.80	0.75002431
0.85	0.77421486	0.85	0.70113336
0.90	0.72420476	0.90	0.65348524
0.95	0.67481728	0.95	0.60771055
1.00	0.62687730	1.00	0.56418957
1.10	0.53730304	1.10	0.48460502
1.20	0.45765528	1.20	0.41512292
1.30	0.38828294	1.30	0.35517588
1.40	0.32863543	1.40	0.30380810
1.50	0.27776957	1.50	0.25995954

Kappa, 0.90 (*cont.*)		Kappa, 1.00 (*cont.*)	
τ	$F(\tau)$	τ	$F(\tau)$
1.60	0.23462009	1.60	0.22260187
1.70	0.19813773	1.70	0.19079750
1.80	0.16735434	1.80	0.16371953
1.90	0.14140869	1.90	0.14065269
2.00	0.11955136	2.00	0.12098536
2.10	0.10113928	2.10	0.10419809
2.20	0.08562557	2.20	0.08985155
2.30	0.07254797	2.30	0.07757490
2.40	0.06151722	2.40	0.06705549
2.50	0.05220632	2.50	0.05802965
2.60	0.04434097	2.60	0.05027499
2.70	0.03769126	2.70	0.04360370
2.80	0.03206452	2.80	0.03785709
2.90	0.02729928	2.90	0.03290078
3.00	0.02326015	3.00	0.02862093
3.10	0.01983355	3.10	0.02492089
3.20	0.01692413	3.20	0.02171849
3.30	0.01445177	3.30	0.01894379
3.40	0.01234911	3.40	0.01653715
3.50	0.01055944	3.50	0.01444764
3.60	0.00903499	3.60	0.01263172
3.70	0.00773549	3.70	0.01105210
3.80	0.00662692	3.80	0.00967679
3.90	0.00568056	3.90	0.00847834
4.00	0.00487213	4.00	0.00743314

Kappa, 1.10		Kappa, 1.20	
τ	$F(\tau)$	τ	$F(\tau)$
0.05	0.00001524	0.05	0.00015136
0.10	0.02008037	0.10	0.05362131
0.15	0.16485130	0.15	0.28538564
0.20	0.40732400	0.20	0.56963938
0.25	0.63907235	0.25	0.78840380
0.30	0.80930594	0.30	0.92036345
0.35	0.91340065	0.35	0.98191851
0.40	0.96363692	0.40	0.99476334
0.45	0.97485249	0.45	0.97653381
0.50	0.95965880	0.50	0.93970627
0.60	0.88530559	0.60	0.84060510
0.70	0.78747780	0.70	0.73421028
0.80	0.68778281	0.80	0.63464248
0.90	0.59522419	0.90	0.54642304

τ	Kappa, 1.10 (*cont.*) $F(\tau)$	τ	Kappa, 1.20 (*cont.*) $F(\tau)$
1.00	0.51289960	1.00	0.47015797
1.10	0.44124565	1.10	0.40496084
1.20	0.37957402	1.20	0.34947719
1.30	0.32679014	1.30	0.30230770
1.40	0.28172201	1.40	0.26217057
1.50	0.24326258	1.50	0.22795093
1.60	0.21042394	1.60	0.19870341
1.70	0.18234979	1.70	0.17363632
1.80	0.15830959	1.80	0.15209041
1.90	0.13768551	1.90	0.13351786
2.00	0.11995734	2.00	0.11746328
2.10	0.10468791	2.10	0.10354750
2.20	0.09150993	2.20	0.09145410
2.30	0.08011459	2.30	0.08091824
2.40	0.07024196	2.40	0.07171758
2.50	0.06167276	2.50	0.06366485
2.60	0.05422170	2.60	0.05660180
2.70	0.04773184	2.70	0.05039431
2.80	0.04206999	2.80	0.04492831
2.90	0.03712278	2.90	0.04010652
3.00	0.03279359	3.00	0.03584572
3.10	0.02899984	3.10	0.03207453
3.20	0.02567079	3.20	0.02873153
3.30	0.02274574	3.30	0.02576377
3.40	0.02017248	3.40	0.02312548
3.50	0.01790605	3.50	0.02077696
3.60	0.01590759	3.60	0.01868377
3.70	0.01414354	3.70	0.01681592
3.80	0.01258479	3.80	0.01514724
3.90	0.01120609	3.90	0.01365487
4.00	0.00998550	4.00	0.01231880
4.10	0.00890390	4.10	0.01112147
4.20	0.00794464	4.20	0.01004746
4.30	0.00709318	4.30	0.00908318
4.40	0.00633679	4.40	0.00821668
4.50	0.00566435	4.50	0.00743739

τ	Kappa, 1.40 $F(\tau)$	τ	Kappa, 1.60 $F(\tau)$
0.05	0.00360864	0.02	0.00000088
0.10	0.20441307	0.04	0.00543957
0.15	0.59414010	0.06	0.07616474
0.20	0.88044006	0.08	0.24992655

Kappa, 1.40 (*cont.*)		Kappa, 1.60 (*cont.*)	
τ	$F(\tau)$	τ	$F(\tau)$
0.25	1.02290360	0.10	0.47116194
0.30	1.06586710	0.12	0.68194478
0.35	1.05127820	0.14	0.85486199
0.40	1.00643580	0.16	0.98397718
0.45	0.94738840	0.18	1.07318150
0.50	0.88318690	0.20	1.12951500
0.60	0.75680232	0.25	1.17136150
0.70	0.64440849	0.30	1.13379430
0.80	0.54901358	0.35	1.06271230
0.90	0.46932237	0.40	0.98069380
1.00	0.40299254	0.45	0.89835155
1.10	0.34769116	0.50	0.82040244
1.20	0.30139719	0.60	0.68365846
1.30	0.26244717	0.70	0.57259344
1.40	0.22949946	0.80	0.48326769
1.50	0.20147925	0.90	0.41120306
1.60	0.17752593	1.00	0.35261848
1.70	0.15694837	1.10	0.30456067
1.80	0.13918892	1.20	0.26477582
1.90	0.12379528	1.30	0.23155026
2.00	0.11039836	1.40	0.20357520
2.10	0.09869530	1.50	0.17984311
2.20	0.08843610	1.60	0.15957126
2.30	0.07941325	1.70	0.14214567
2.40	0.07145360	1.80	0.12708012
2.50	0.06441189	1.90	0.11398607
2.60	0.05816571	2.00	0.10255030
2.70	0.05261141	2.10	0.09251828
2.80	0.04766083	2.20	0.08368157
2.90	0.04323869	2.30	0.07586826
3.00	0.03928046	2.40	0.06893563
3.10	0.03573063	2.50	0.06276440
3.20	0.03254124	2.60	0.05725437
3.30	0.02967075	2.70	0.05232088
3.40	0.02708308	2.80	0.04789200
3.50	0.02474676	2.90	0.04390638
3.60	0.02263432	3.00	0.04031142
3.70	0.02072164	3.10	0.03706184
3.80	0.01898757	3.20	0.03411851
3.90	0.01741347	3.30	0.03144749
4.00	0.01598286	3.40	0.02901923
4.10	0.01468121	3.60	0.02479095

Kappa, 1.40 (cont.)		Kappa, 1.60 (cont.)	
τ	$F(\tau)$	τ	$F(\tau)$
4.20	0.01349562	3.80	0.02126290
4.30	0.01241461	4.00	0.01830252
4.40	0.01142799	4.20	0.01580583
4.50	0.01052668	4.40	0.01369049

Kappa, 1.80		Kappa, 2.00	
τ	$F(\tau)$	τ	$F(\tau)$
0.02	0.00004054	0.02	0.00060974
0.04	0.03197026	0.04	0.11111402
0.06	0.22643367	0.06	0.48332666
0.08	0.52888149	0.08	0.88521996
0.10	0.81361016	0.10	1.17746710
0.12	1.02888330	0.12	1.35195250
0.14	1.17175930	0.14	1.43755390
0.16	1.25559480	0.16	1.46354290
0.18	1.29575900	0.18	1.45178900
0.20	1.30519300	0.20	1.41714570
0.25	1.25212930	0.25	1.28586220
0.30	1.15222210	0.30	1.14123810
0.35	1.04290140	0.35	1.00746300
0.40	0.93848040	0.40	0.89040751
0.45	0.84377785	0.45	0.78993385
0.50	0.75976068	0.50	0.70413068
0.60	0.62112821	0.60	0.56782591
0.70	0.51436522	0.70	0.46643222
0.80	0.43133578	0.80	0.38934212
0.90	0.36584693	0.90	0.32947685
1.00	0.31343865	1.00	0.28209478
1.10	0.27092244	1.10	0.24396022
1.20	0.23600035	1.20	0.21281590
1.30	0.20699398	1.30	0.18705244
1.40	0.18265917	1.40	0.16549870
1.50	0.16205984	1.50	0.14728637
1.60	0.14448183	1.60	0.13176074
1.70	0.12937301	1.70	0.11842051
1.80	0.11630130	1.80	0.10687653
1.90	0.10492480	1.90	0.09682288
2.00	0.09497007	2.00	0.08801632
2.10	0.08621645	2.10	0.08026145
2.20	0.07848420	2.20	0.07339976
2.30	0.07162579	2.30	0.06730158
2.40	0.06551917	2.40	0.06185997

Kappa, 1.80 (cont.)		Kappa, 2.00 (cont.)	
τ	$F(\tau)$	τ	$F(\tau)$
2.50	0.06006274	2.50	0.05698607
2.60	0.05517129	2.60	0.05260557
2.70	0.05077299	2.70	0.04865588
2.80	0.04680695	2.80	0.04508404
2.90	0.04322121	2.90	0.04184490
3.00	0.03997133	3.00	0.03889988
3.10	0.03701903	3.10	0.03621577
3.20	0.03433126	3.20	0.03376389
3.30	0.03187931	3.30	0.03151935
3.40	0.02963822	3.40	0.02946050
3.60	0.02570387	3.60	0.02582641
3.80	0.02238351	3.80	0.02273605
4.00	0.01956451	4.00	0.02009159
4.20	0.01715823	4.20	0.01781592
4.40	0.01509419	4.40	0.01584760

Kappa, 2.50			
τ	$F(\tau)$	τ	$F(\tau)$
0.01	0.00003491	1.10	0.19532792
0.02	0.03674250	1.20	0.17076425
0.03	0.28737946	1.30	0.15057725
0.04	0.70698777	1.40	0.13376781
0.05	1.12405520	1.50	0.11960986
0.06	1.45527340	1.60	0.10756487
0.07	1.68762140	1.70	0.09722613
0.08	1.83517030	1.80	0.08828151
0.10	1.95271760	1.90	0.08048800
0.12	1.93325720	2.00	0.07365402
0.14	1.85013120	2.10	0.06762671
0.16	1.74127320	2.20	0.06228286
0.18	1.62556100	2.30	0.05752218
0.20	1.51210270	2.40	0.05326235
0.25	1.25958960	2.50	0.04943523
0.30	1.05756610	2.60	0.04598401
0.35	0.89846695	2.80	0.04002593
0.40	0.77242557	3.00	0.03508759
0.45	0.67136010	3.20	0.03095001
0.50	0.58923222	3.40	0.02745031
0.60	0.46529486	3.60	0.02446526
0.70	0.37748868	3.80	0.02190013
0.80	0.31287848	4.00	0.01968108
0.90	0.26384519	4.20	0.01774985
1.00	0.22567582	4.40	0.01605995

Moments of $F(\tau)$ for the Compartmental System

For a system of n equal compartments, of magnitude S, which are being washed out at a uniform rate ρ we have from eq. 25, Ch. 9

$$\frac{a_i(t)}{a(0)} = \frac{(\rho t/s)^{n-1}e^{-\rho t/S}}{(n-1)!}$$

In terms of $F(\tau)$ and τ this becomes

$$F(\tau) = \frac{n(n\tau)^{n-1}e^{-n\tau}}{(n-1)!}$$

since the system volume on which τ is determined is n times S. We define the ith moment about the origin (eq. 37, Ch. 9) as

$$m_i = \int_0^\infty \tau^i \frac{(n\tau)^{n-1}e^{-n\tau}}{(n-1)!}\, d\tau$$

Integrating and simplifying yield

$$m_i = \frac{(n+i-1)(n+i-2)\cdots n}{n^i}$$

For the first moment $m_1 = 1$,

for the second moment $m_2 = (n+1)/n$,

and the third $m_3 = \dfrac{(n+2)(n+1)}{n^2}$.

As $n \to \infty$ all moments $\to 1$. Thus for large n the distribution tends to become highly peaked at $\bar{x} = 1$.

References

1. Daniel E. Feldman, Henry T. Yost Jr., and Bruce B. Benson, "Oxygen Isotope Fractionation in Reactions Catalyzed by Enzymes," *Science* **129**: 146–147, 1959.

2. Leslie Corsa, Jr., John M. Olney, Jr., Richard W. Steenburg, Margaret R. Ball, and Francis D. Moore, "The Measurement of Exchangeable Potassium in Man by Isotope Dilution," *J. Clin. Invest.* **29**: 1280–1295, 1950.

3. L. Brillouin, *Science and Information Theory*, Academic Press, New York, 1956.

4. C. W. Sheppard, and A. S. Householder, "The Mathematical Basis of the Interpretation of Tracer Experiments in Closed Steady-State Systems," *J. Appl. Physics* **22**: 510–520, 1951.

5. Ernest A. Pinson, "Water Exchanges and Barriers as Studied by the Use of Hydrogen Isotopes," *Physiol. Revs.* **32**: 123–134, 1952.

6. C. W. Sheppard, "Compound Interest Laws and Disappearance of Tracers from the Circulation," *Circulation Research* **5**: 220–222, 1957.

7. R. B. Dean, T. R. Noonan, L. Haege, and W. O. Fenn, "Permeability of Erythrocytes to Radioactive Potassium," *J. Gen. Physiol.* **24**: 353–365, 1940.

8. C. W. Sheppard and W. R. Martin, "Cation Exchange Between Cells and Plasma of Mammalian Blood; I: Methods and Application to Potassium Exchange in Human Blood," *J. Gen. Physiol.* **33** 703–722, 1950.

9. R. B. Duffield and M. Calvin, "The Stability of Chelate Compounds: III. Exchange Reactions of Copper Chelate Compounds," *J. Am. Chem. Soc.* **68**: 557–561, 1946.

10. D. B. Zilversmit, and M. L. Shore, "A Hydrodynamic Model of Isotope Distribution in Living Organisms," *Nucleonics* **10**: No. 10: 32–34, 1952.

11. C. W. Sheppard, "The Theory of the Study of Transfers Within a Multi-Compartment System Using Isotopic Tracers," *J. Appl. Physics* **19** 70–76, 1948.

12. S. M. Skinner, R. E. Clark, N. Baker, and R. A. Shipley, "Complete Solution of the Three-Compartment Model in Steady State After Single Injection of Radioactive Tracer," *Am. J. Physiol.* **196**: 238–244, 1959.

13. J. S. Robertson, D. C. Tosteson, and J. L. Gamble, "The Determination of Exchange Rates in Three-Compartment Steady-State Closed Systems Through the Use of Tracers," *J. Lab. Clin. Med.* **49**: 497–503, 1957.

14. A. Gellhorn, M. Merrell, and R. M. Rankin, "The Rate of Transcapillary Exchange of Sodium in Normal and Shocked Dogs," *Am. J. Physiol.* **142**: 407–427, 1944.

15. W. E. Cohn, and A. M. Brues, "Metabolism of Tissue Cultures: III. A Method for Measuring the Permeability of Tissue Cells to Solutes," *J. Gen. Physiol.* **28**: 449–461, 1945.

16. L. E. Dickson, *First Course in the Theory of Equations*, Wiley, New York, 1922.

17. G. Hetenyi, Jr., A. M. Rappaport, and G. A. Wrenshall, "The Validity of Rates of Glucose Appearance in the Dog Calculated by the Method of Successive Tracer Injections; I. Effects of Surgical Hepatectomy, Evisceration, and Order of Tracer Injections," *Can. J. Biochem. Physiol.* **39**: 225–236, 1961.

18. Howard Gest, Martin D. Kamen, and John M. Reiner, "The Theory of Isotope Dilution," *Arch. Biochem. Biophys.* **12**: 273–281, 1947.

19. John M. Reiner, "The Study of Metabolic Turnover Rates by Means of Isotopic Tracers: I. Fundamental Relations; II. Turnover in a Simple Reaction System," *Arch. Biochem. Biophys.* **46**: 53–99, 1953.

20. J. S. Robertson, "Theory and Use of Tracers in Determining Transfer Rates in Biological Systems," *Physiol. Revs.* **37**: 133–154, 1957.

21. J. A. Russell, "The Use of Isotopic Tracers in Estimating Rates of Metabolic Reactions," *Perspectives Biol. Med.* **1**: 138–173, 1958.

22. A. K. Solomon, "The Kinetics of Biological Processes: Special Problems Connected with the Use of Tracers," *Advances in Biol. and Med. Physics* **3**: 65–97, 1953.

23. A. K. Solomon, "Equations for Tracer Experiments," *J. Clin. Invest.* **28**: 1297–1307, 1949.

24. D. B. Zilversmit, "The Design and Analysis of Isotope Experiments," *Am. J. Med.* **29**: 832–848, 1960.

25. S. Aronoff, "Techniques of Radiobiochemistry," Iowa State College Press, Ames, Iowa, 1956.

26. C. L. Comar, *Radioisotopes in Biology and Agriculture*, McGraw-Hill, New York, 1955.

27. G. Hevesy, *Radioactive Indicators: Their Application in Biochemistry Animal Physiology and Pathology*, Interscience, New York, 1948.

28. W. E. Siri, *Isotopic Tracers and Nuclear Radiations*, McGraw-Hill, New York, 1949.

29. H. E. Hart, "Analysis of Tracer Experiments in Non-Conservative Steady-State Systems," *Bull. Math. Biophys.* **17**: 87–94, 1955.

30. H. E. Hart, "Analysis of Tracer Experiments: II. Non-Conservative Non-Steady-State Systems," *Bull. Math. Biophys.* **19**: 61–72, 1957.

31. H. E. Hart, "Analysis of Tracer Experiments: III. Homeostatic Mechanisms of Fluid-Flow Systems," *Bull. Math. Biophys.* **20**: 281–287, 1958.

32. H. D. Landahl, "Note on the Interpretation of Tracer Experiments in Biological Systems," *Bull. Math. Biophys.* **16**: 151–154, 1954.

33. Aldo Rescigno, "A Contribution to the Theory of Tracer Methods," *Biochim. et Biophys. Acta* **15**: 340–344, 1954.

34. Aldo Rescigno, "A Contribution to the Theory of Tracer Methods, Part II," *Biochim. et Biophys. Acta* **21**: 111–116, 1956.

35. J. L. Stephenson, "Theory of Transport in Linear Biological Systems: II. Multiflux Problems," *Bull. Math. Biophys.* **22**: 113–138, 1960.

36. Gerald A. Wrenshall, "A Working Basis for the Tracer Measurement of Transfer Rates of a Metabolic Factor in Biological Systems Containing Compartments Whose Contents Do Not Intermix Rapidly," *Can. J. Biochem. and Physiol.* **33**: 909–925, 1955.

37. T. Teorell, "Kinetics of Distribution of Substances Administered to the Body: I. The Extravascular Modes of Administration; II. The Intravascular Modes of Administration," *Arch. intern. pharmacodyn.* **57**: 205–240, 1937.

38. S. P. Thompson, *Calculus Made Easy*, Macmillan, London, 1957.

39. M. F. Gardner, and J. L. Barnes, *Transients in Linear Systems*, Vol. 1, Wiley, New York, 1942.

40. Bateman Manuscript Project, A. Erdelyi, (Editor), *Tables of Integral Transforms:* I; McGraw-Hill, New York, 1954.

41. L. R. Ford, *Differential Equations*, McGraw-Hill, New York, 1933.

42. D. B. Zilversmit, C. Entenman, M. C. Fishler, "On the Calculation of Turnover Time and Turnover Rate from Experiments Involving the Use of Labeling Agents," *J. Gen. Physiol.* **26**: 325–331, 1943.

43. C. W. Sheppard, W. R. Martin, and G. E. Beyl, "Cation Exchange Between Cells and Plasma of Mammalian Blood: II. Sodium and Potassium Exchange in the Sheep, Cow, Dog, and Man and the Effect of Varying the Plasma Potassium Concentration," *J. Gen. Physiol.* **34**: 411–429, 1951.

44. Max Kleiber, "Use of P^{32} as a Tracer in Research on Metabolism and Food Utilization in Intact Dairy Cows," Pages 99–121 in *A Conference on the Use of Isotopes in Plant and Animal Research, USAEC* Report TID 5098, 1953.

45. Harry Schachter, "Direct Versus Tracer Measurement of Transfer Rates in a Hydrodynamic System Containing a Compartment Whose Contents Do Not Intermix Rapidly," *Can. J. Biochem. and Physiol.* **33**: 940–947, 1955.

46. C. Garavaglia, C. Polvani, and R. Silvestrini, *A Collection of Curves Obtained with a Hydrodynamic Model Simulating Some Schemes of Biological Experiments Carried out with Tracers*, Report No. 60, Centro Informazioni Studi Esperienze Milan, 1958.

47. A. A. Plentl, and M. J. Gray, "Hydrodynamic Model of A 3-Compartment Catenary System with Exchanging End Compartments," *Proc. Soc. Exp. Biol. Med.* **87**: 595–600, 1954.

48. J. R. MacDonald, E. G. Perry, L. L. Madison, and D. W. Seldin, "An Electrical Analogue for Analysis of Tracer Distribution Kinetics in Biological Systems," *Radiation Research* **6**: 585–601, 1957.

49. J. R. Macdonald, "New Integrating Circuit and Electrical Analog for Transient Diffusion and Flow," *Rev. Sci. Instr.* **28**: 924–926, 1957.

50. E. Rollinson, and J. Rotblat, "An Electrical Analogue of the Metabolism of Iodine in the Human Body," *Brit. J. Radiol.* **28**: 191–198, 1955.

51. G. L. Brownell, R. V. Cavicchi, K. E. Perry, "An Electrical Analog for Analysis of Compartmental Biological Systems," *Rev. Sci. Instr.* **24**: 704, 1953.

52. G. A. Korn, and T. M. Korn, *Electronic Analog Computers*, McGraw-Hill, New York, 1956.

53. J. B. Stanbury, G. L. Brownell, D. S. Riggs, H. Perinetti, J. Itoiz, and E. B. Del Castillo, *Endemic Goiter: The Adaptation of Man to Iodine Deficiency*, Harvard University Press, Cambridge, Mass., 1954.

54. C. W. Sheppard and G. E. Beyl, "Cation Exchange in Mammalian Erythrocytes: III. The Prolytic Effect of X-Rays on Human Cells." *J. Gen. Physiol.* **34**: 691–704, 1951.

55. O. Taussky, "Ordinary Differential Equations." In E. U. Condon and H. Odishaw (Eds.), *Handbook of Physics*, McGraw-Hill, New York, 1958.

56. *IBM 650 Magnetic Drum Data-Processing Machine*, International Business Machines Corp., 590 Madison Ave., New York, 1955.

57. R. V. Andree, *Programming the IBM 650 Magnetic Drum Computer and Data-Processing Machine*, Holt, Rinehart and Winston, New York, 1958.

58. *FOR TRANSIT Automatic Coding System for the IBM 650 Data Processing System* (IBM Reference Manual), International Business Machines Corp., 590 Madison Ave., New York, 1959.

59. C. Edwards, and E. J. Harris, "Do Tracers Measure Fluxes?", *Nature* **175**: 262, 1955.

60. E. U. Condon, "Vector Analysis." In E. U. Condon and H. Odishaw (Eds.), *Handbook of Physics*, McGraw-Hill, New York, 1958.

61. H. S. Carslaw, and J. C. Jaeger, *Conduction of Heat in Solids*, Clarendon Press, Oxford, 1948.

62. V. Paschkis and J. W. Hlinka, "Electric Analogy Studies of the Transient Behavior of Heat Exchangers," *Trans. N. Y. Acad. Sci.* **19**: 714–730, 1957.

63. E. J. Harris and G. P. Burn, "The Transfer of Sodium and Potassium Ions Between Muscle and the Surrounding Medium," *Trans. Faraday Soc.* **45**: 508–28, 1949.

64. Herman Branson, "The Use of Isotopes in an Integral Equation Description of Metabolizing Systems," *Cold Spring Harbor Symposia Quant. Biol.* **13**: 35–42, 1948.

65. Herman Branson, "Metabolic Pathways from Tracer Experiments," *Arch. Biochem. Biophys.* **36**: 60–70, 1952.

66. J. Z. Hearon, "A Note on the Integral Equation Description of Metabolizing Systems," *Bull. Math. Biophys.* **15**: 269–276, 1953.

67. Robert A. Wijsman, "A Critical Investigation of the Integral Description of Metabolizing Systems," *Bull. Math. Biophys.* **15**: 261–268, 1953.

68. J. L. Stephenson, "Theory of Measurement of Blood Flow by Dye Dilution Technique," *IRE Trans. on Med. Electronics* **PGME-9,** 82–88, 1958.

69. N. J. Nadler, "A Theory for a Quantitative Investigation of the Turnover of Iodine in the Thyroid Gland," *Endocrinology* **62**: 758–767, 1958.

70. Hans H. Ussing, "Some Aspects of the Application of Tracers in Permeability Studies." *Advances in Enzymol.* **13**: 21–65, 1952.

71. L. F. Nims, "Membranes, Tagged Components and Membrane Transfer Coefficient," *Yale J. Biol. and Med.* **31**: 373–386, 1959.

72. P. F. Hahn, J. F. Ross, W. F. Bale, W. M. Balfour, and G. H. Whipple, "Red Cell and Plasma Volumes, (Circulating and Total) as Determined by Radio Iron and by Dye," *J. Exp. Med.* **75**: 221–232, 1942.

73. P. Dow, "Estimations of Cardiac Output and Central Blood Volume by Dye Dilution," *Physiol. Revs.* **36**: 77–102, 1956.

74. Sir Geoffrey Taylor, "The Dispersion of Soluble Matter in Solvent Flowing Slowly through a Tube," *Proc. Roy. Soc.* (London), Ser. A **219**: 186–203, 1953.

75. H. Sherman, R. C. Schlant, W. L. Kraus, and C. B. Moore, "A Figure of Merit for Catheter Sampling Systems," *Circulation Research* **7**: 303–313, 1959.

76. C. W. Sheppard, M. P. Jones, and B. L. Couch, "Effect of Catheter Sampling on the Shape of Indicator-Dilution Curves. Mean Concentration versus Mean Flux of Outflowing Dye," *Circulation Research* **7**: 895–906, 1959.

77. H. H. Rossi, S. H. Powers, and B. Dwork, "Measurement of Flow in Straight Tubes by Means of the Dilution Technique," *Am. J. Physiol.* **173**: 103–108, 1953.

78. W. F. Hamilton, J. W. Moore, J. M. Kinsman, and R. G. Spurling, "Studies on the Circulation: IV. Further analysis of the injection method and of changes in hemodynamics under physiological and pathological conditions, *Am. J. Physiol.* **99**: 534–551, 1931.

79. P. Meier and K. L. Zierler, "On the Theory of the Indicator-Dilution Method for Measurement of Blood Flow and Volume," *J. Appl. Physiol* **6**: 731–744, 1954.

80. P. Dow, P. F. Hahn, and W. F. Hamilton, "The Simultaneous Transport of T-1824 and Radioactive Red Cells through the Heart and Lungs," *Am. J. Physiol.* **147**: 493–499, 1946.

81. P. F. Hahn, W. D. Donald, R. C. Grier, Jr., "The Physiological Bilaterality of the Portal Circulation," *Am. J. Physiol.* **143**: 105–107, 1945.

82. C. W. Sheppard and L. J. Savage, "The Random Walk Problem in Relation to the Physiology of Circulatory Mixing," *Phys. Rev.* **83**: 489, 1951.

83. C. W. Sheppard, "Synthesis of Dye Dilution Curves," *Am. J. Physiol.* **171**: 767, 1952.

84. C. W. Sheppard, "Mathematical Considerations of Indicator Dilution Techniques," *Minn. Med.* **37**: 93–104, 1954.

85. S. Chandrasekhar, "Stochastic Problems in Physics and Astronomy," *Revs. Modern Phys.* **15**: 1–89, 1943.

86. E. V. Newman, M. Merrell, A. Genecin, C. Monge, W. R. Milnor, and W. P. McKeever, "The Dye Dilution Method for Describing the Central Circulation. An Analysis of factors Shaping the Time Concentration Curves," *Circulation* **4**: 735–746, 1951.

87. W. Feller, *An Introduction to Probability Theory and Its Applications*, Wiley, New York, 1950.

88. H. F. Stegall, R. B. Swann, and W. D. Collings, "A Method for Analyzing An Organ's Internal Circulations," *Texas Repts. Biol. and Med.* **18**: 558–565, 1960.

89. C. W. Sheppard, "An Electromathematical Theory of Circulatory Mixing Transients," *Proc. First Nat. Biophysics Conf.*, Yale University Press, New Haven, 1959, pp. 476–492.

90. D. Parrish, D. T. Hayden, W. Garrett, and R. L. Huff, "Analog Computer Analysis of Flow Characteristics and Volume of the Pulmonary Vascular Bed," *Circulation Research* **7**: 746–752, 1959.

91. J. L. Stephenson, "Theory of Measurement of Blood Flow by the Dilution of an Indicator," *Bull. Math. Biophys.* **10**: 117–121, 1948.

92. D. T. Hayden, W. Garrett, and P. Jordan, "Evaluation of Mitral Insufficiency in Dogs by Electronic Analog Simulation of Radioisotope Dilution Data," *Circulation Research* **6**: 77–82, 1958.

93. J. A. McClure, W. W. Lacy, P. Latimer, and E. V. Newman, "Indicator Dilution in an Atrioventricular System with Competent or Incompetent Valves," *Circulation Research* **7**: 794–806, 1959.

94. G. W. Schmidt, "A Mathematical Theory of Capillary Exchange as a Function of Tissue Structure," *Bull. Math. Biophys.* **14**: 229–263, 1952.

95. R. Macey, "A Probabilistic Approach to Some Problems in Blood-Tissue Exchange," *Bull. Math. Biophys.* **18**: 205–217, 1956.

96. J. J. Blum, "Concentration Profiles in and Around Capillaries," *Am. J. Physiol.* **198**: 991–998, 1960.

97. W. C. Sangren and C. W. Sheppard, "A Mathematical Derivation of the Exchange of a Labeled Substance Between a Liquid Flowing in a Vessel and an External Compartment," *Bull. Math. Biophys.* **15**: 387–394, 1953.

98. H. D. Landahl, "Transient Phenomena in Capillary Exchange," *Bull. Math. Biophys.* **16**: 55–58, 1954.

99. C. W. Sheppard, R. R. Overman, W. S. Wilde, and W. C. Sangren, "The Disappearance of K^{42} from the Nonuniformly Mixed Circulation Pool in Dogs," *Circulation Research* **1**: 284–297, 1953.

100. C. W. Sheppard and W. C. Sangren, "Transcapillary Movement of an Isotope from a Nonuniformly Mixed Circulatory Pool," *Federation Proc.* **11**: 147, 1952.

101. D. A. K. Black, H. E. F. Davies, and E. W. Emery, "The Disposal of Radioactive Potassium Injected Intravenously," *Lancet* **268**: 1097–99, 1955.

102. L. A. Sapirstein, "Fractionation of the Cardiac Output of Rats with Isotopic Potassium," *Circulation Research* **4**: 689–692, 1956.

103. A. C. Dornhorst, "The Interpretation of Red Cell Survival Curves," *Blood* **6**: 1284–92, 1951.

104. B. L. Strehler and A. S. Mildvan, "General Theory of Mortality and Aging," *Science* **132**: 14–21, 1960.

105. C. G. Craddock, Jr., S. Perry, and J.S. Lawrence, "The Dynamics of Leukopenia and Leukocytosis," *Ann. Internal Med.* **52**: 281–294, 1960.

106. P. F. Hahn and N. Wiener, "Mathematical Relationships of Possible Significance in the Study of Human Leukemia," *Federation Proc.* **10**: 357, 1951.

107. A. D. Michal, *Matrix and Tensor Calculus*, Wiley, New York, 1947.

108. Per-Erik E. Bergner, "Tracer Dynamics: I. A Tentative Approach and Definition of Fundamental Concepts," *J. Theo. Biol.* (In press.)

109. V. N. Faddeeva, *Computational Methods of Linear Algebra*, Dover Publications, New York, 1959.

110. M. Berman and R. Schoenfeld, "Invariants in Experimental Data on Linear Kinetics and the Formulation of Models," *J. Appl. Physics*, **27**: 1361–1370, 1956.

111. C. Lanczos, *Applied Analysis*, Prentice-Hall, Englewood Cliffs, N. J., 1956.

112. A. M. Mood, *Introduction to the Theory of Statistics*, McGraw-Hill, New York, 1950.

113. E. Whittaker and G. Robinson, *The Calculus of Observations*, Blackie and Son, London, 1924.

114. W. E. Deming, *Statistical Adjustment of Data*, Wiley, New York, 1943.

115. B. H. Worsley, D. B. W. Reid and Louis C. Lax, "Error Estimation in Transfer Rates of Plasma Constituents," *Proceedings of the Second Conference of the Computing and Data Processing Society of Canada*, 158–174, Toronto, 1960.

116. A. S. Householder, *On Prony's Method of Fitting Exponential Decay Curves and Multiple-Hit Survival Curves*. USAEC Report No. ORNL 455, Series B, Oak Ridge National Laboratory, Oak Ridge, 1949.

117. R. G. Cornell, *A New Estimation Procedure for Linear Combinations of Exponentials*. Ph.D. Thesis, Virginia Polytechnic Institute, Blacksburg, Va., 1956. Obtainable from University Microfilms, Ann Arbor, Mich.

118. W. Perl, "A Method for Curve-Fitting by Exponential Functions," *Int. J. Appl. Radiation and Isotopes* **8**: 211–222, 1960.

119. J. Cornfield, J. Steinfeld, and S. W. Greenhouse, "Models for the Interpretation of Experiments Using Tracer Compounds," *Biometrics* **16**: 212–234, 1960.

120. E. H. Kennard, *Kinetic Theory of Gases*, McGraw-Hill, New York, 1938.

Index

Absolute amounts of a tracer, 9
Absolute exchange, 218
Absolute value, 135
Abundance ratios, 9, 19, 31
Access to compartments, 30
Adder, 101, 105, 231
Additivity of functions and transforms, 50
Addressable storage positions, 129
Addresses, 130
Age concept, 160, 162, 181, 185
Alphabetic data, 130
Alphabetic symbols, 130
Analog computation, 4, 103
Analog computer, 97, 229
Analog computers, 92
Analog methods, 14
Analog simulation, 89
Analog solution, 106
Analog systems, 14
Analogy, electronic, 103
  hydrodynamical, 103
Analytical determination, 3
Ancillary information, 33
Anisotropic system, 155
Aorta, 176
Aortic outflow, 228
Aperiodic phase, 215
Applied mathematics, 50
Approximate dispersion functions, 208
Approximate values, 134
Approximation methods, 79, 81

Arithmetic mean, 180, 229, 244
Arsenic-76, 3
Arsenic, tissue levels, 3
  trypanocidal effect of, 3
Arterio-venous differences, 103, 221, 223, 224, 227
Asymptote, 63
Asymptotic expressions, 201

Backflow, 216, 217, 219, 230
  postulated, 225
Background noise, 224
Bacterial colonies, 185
Balance adjustment in amplifier, 107
Beds, parallel-connected, 204
Bessel function, 221
Best fit, 242
Best value, 248, 249
Biasing effect, 29
Binary digits, 4
Binomial law, 193
Biological half-life, 7, 103, 112
Birth function, 232
Block diagrams, 101
Blood, 182
Blood flow, 176, 227
Body fluids, 6
Body water, 6
Bolus, 166, 178, 223
Bolus response, 175
Bolus type, 182
Boundaries, 153